Solar Photovoltaic Power Optimization

This book focuses on the rapidly maturing solar photovoltaic (PV) industry, which is achieving an ever-increasing share of U.S. and global power production.

There is a growing need for all stakeholders – owners, maintenance technicians, utilities, and installers – to fully understand the operations and maintenance of PV systems, and how to monitor and diagnose systems post installation. Recognizing this need, this book covers monitoring and diagnostic techniques and technologies, including how to identify the causes of poor performance, and measure and verify power production. Drawing on global case studies, it details how to achieve optimal PV power output in the field through an overview of basic electrical, the solar PV module and Balance of System, and processes and software for monitoring, measurement, and verification. It also provides an overview of the North American Board of Certified Energy Practitioner's (NABCEP) new PV System Inspector credential, which will be outlined in the final chapter.

Equipping the reader with the knowledge and confidence required to maximize the output of solar PV installations, *Solar Photovoltaic Power Optimization* will be an essential resource for PV practitioners and students.

Michael Ginsberg is a LEED AP O+M, CEM, and NABCEP PV Associate, holds an MS in Sustainability Management from Columbia University, and is a Doctor of Engineering Science candidate at Columbia University specializing in solar energy integration into the electrical grid. In his work with the U.S. Department of State, Michael has performed technical analyses of large-scale solar installs on U.S. compounds worldwide, and trained nearly 1,000 engineers and technicians in renewable energy and building systems at U.S. embassies in West Africa, South America, the Middle East, and Europe.

Solar Photovoltaic Power Optimization

Enhancing System Performance through Operations, Measurement, and Verification

Michael Ginsberg

Routledge
Taylor & Francis Group
LONDON AND NEW YORK

earthscan
from Routledge

First published 2020
by Routledge
2 Park Square, Milton Park, Abingdon, Oxon OX14 4RN

and by Routledge
52 Vanderbilt Avenue, New York, NY 10017

Routledge is an imprint of the Taylor & Francis Group, an informa business

British Library Cataloguing-in-Publication Data
A catalogue record for this book is available from the British Library

Library of Congress Cataloging-in-Publication Data
A catalog record for this book has been requested

ISBN: 978-0-8153-9859-2 (hbk)
ISBN: 978-0-8153-9867-7 (pbk)
ISBN: 978-1-351-17204-2 (ebk)

Typeset in Times New Roman
by Apex CoVantage, LLC

This book is dedicated to my grandparents Michael Roach, Barbara Parker, Cyma Satell, Howard Ginsberg, and Ed Satell, who set me on the road to solar energy, and helped me to realize my dreams.

Contents

Acknowledgments viii
Preface x

1 The need for solar photovoltaic system operations and maintenance, and measurement and verification 1

2 Electricity and solar cell and system fundamentals 19

3 Tools and methodology for maintenance and troubleshooting 55

4 Tools and methodology for monitoring, measurement, and verification 119

5 Conclusion: the growing market for PV system inspectors 131

Glossary 140
Index 147

Acknowledgments

I am thankful to a number of individuals who helped shape this book. Richard Driscoll provided exemplary summaries of a variety of key sections, from the history of solar energy to machine learning. Krystyna Larkham's keen editorial eye and organizational prowess were indispensable. My uncle Jeffrey Ginsberg's research informed the analysis of the solar photovoltaic (PV) system.

I am grateful to Brooke Snyder for her research on diagnostic techniques. Thanks to Arthur Hernaez for his research on the theoretical aspects of solar cell failure. I appreciate esteemed PV code expert John Wiles' review and feedback. Thanks to leading PV trainer and developer Sean White for his guidance, and to my doctoral advisor, Dr. Vasilis Fthenakis, for imparting his experience and profound insights into solar PV systems.

Thanks to my friends and colleagues: Susan Gillum Smith, training specialist in the Department of State Bureau of Overseas Building Operations, who guided me for the past decade as a trainer and communicator, and Tim Stufft and Jose-Aponte Aquino, with whom I have seen the world while training and assessing building systems.

The work and programs of the U.S. Department of Energy's National Renewable Energy Laboratory (NREL) are unparalleled. I appreciate the hard work of the bright researchers who provide the basis for our modern understanding of PV systems, chief among whom is NREL Senior Research Fellow Emeritus Dr. Arthur Nozik, and to my great fortune, a guide on my academic journey. Thank you to Electrical Engineering Professor and Director of Distributed Resource Integration at Consolidated Edison (Con Ed) Damian Sciano for helping me to understand the fine details of grid integration.

A huge thanks to my grandfather Michael Roach, a solar energy and microgrid expert. Thanks to my parents for nurturing in me a curiosity for the world and a passion for making a positive contribution. Above all,

I am grateful to my husband, Jeffrey Hanley, for his tireless support of my ambitions.

Solar energy has matured from its infancy into adulthood. I hope this book will help hasten the widespread adoption of solar energy that is needed for the coming global energy transformation.

Preface

Most economic analyses of solar **photovoltaic (PV)** systems, particularly those that compare PV with other energy sources, gloss over **operations and maintenance (O+M)**. This is for good reason – the ongoing costs of harnessing solar energy are minimal in comparison to sources like fossil fuel plants or even wind and water. However, O+M for PV is very different than for other energy sources, and as PV grows in importance, preventable system failures can quickly erode economic return. This can in turn reinforce outdated preconceptions about the high cost of solar and slow the growth of the sector. Sound O+M skills, knowledge, and best practices are vital for everyone from engineers to investors to ensure the industry continues to develop and mature.

As PV becomes mainstream, greater understanding of system components and how they operate, and fail, will be critical for successful adoption. Facilities need to understand the unique intricacies of PV systems to ensure their continued operation, and not rely on existing knowledge of O+M for conventional power sources, such as diesel and gas generators.

The primary audience for this book is technicians, inspectors, engineers, and permitting authorities engaged in solar PV systems on a daily basis, but widespread acceptance of PV as an energy source will require a broader understanding of the technical requirements of operating and maintaining PV systems. Thus, it is hoped that this book, and the deeper understanding of PV systems herein, will also be useful to facility managers, project developers, and even residential DIYers.

It is my philosophy that learning should be fun, especially when it comes to O+M, where the existing volumes of dense instructional manuals that currently serve the industry sit in offices collecting dust. It is time to breathe fresh air and life into O+M, so that PV systems can return the value that investors intended. For example, Chapter 3 of this book analyzes the O+M of each component in a PV system to show the *why* behind specific diagnostic techniques and solutions, before going on to provide analysis and schedule templates that you can use on your own system.

We are at a critical juncture for solar energy. This book aims to provide an intuitive understanding of PV systems that is both tailored to industry professionals and accessible to the public. Accompanied by hands-on experience in the field, this book will equip the reader with the knowledge to minimize PV system downtime and extend equipment life.

Outline

In Chapter 1, we review the history of PV to understand its remarkable transformation from a niche luxury item to one of the fastest and most heavily invested in energy sources by new installations in the world, attracting nearly $160 billion of capital in 2017.[1] In Chapter 2, we review basic electrical concepts and PV system components and their function. In Chapter 3, following a PV-specific review of safety, we determine O+M strategies for each component through a failure modes and effects analysis. In Chapter 4, we discuss the existing and emerging measurement and verification (M+V) tools and techniques for PV systems, including an overview of relevant machine learning techniques. Chapter 5 rounds out the book with forecasts for the future and a summary of the new **North American Board of Certified Energy Practitioners (NABCEP)** credential, **PV System Inspector (PVSI™)**, complete with sample test questions.

Note

1 Globally, new solar power plants added almost 35% to new power generating capacity in 2017. https://phys.org/news/2018-12-globally-solar-power-added-capacity.html. December 18, 2018.

1 The need for solar photovoltaic system operations and maintenance, and measurement and verification

History of the solar photovoltaic industry

The solar photovoltaic (PV) industry evolved in three stages. First, the fundamental science underpinning the industry was established in the 1800s and early 1900s. Next, the silicon PV cell was developed in the 1950s. Finally, low-cost manufacturing methods enabled solar PV to achieve grid parity, or cost equivalence, with conventional fossil fuel generators, making it a viable choice for investors and energy consumers today.

Stage 1: establishing the science

In 1839, 19-year-old Alexandre Edmond Becquerel, who later went on to become a world-famous French physicist, constructed the world's first PV cell. The cell used silver chloride in a liquid acid solution and produced an electric current when light was shone on it, a reaction which is sometimes called the "Becquerel effect." In 1876, William Grylls Adams and Richard Evans Day discovered that the PV effect could be produced in a solid, namely, solidified selenium. Soon after, in 1883, American inventor Charles Fritts coated selenium with a layer of gold, which achieved a power conversion efficiency of 1–2%.[1] In 1905, Albert Einstein published a seminal paper on the photoelectric effect using quantum theory. Eleven years later, Robert Millikan's experiments demonstrated the validity of Einstein's theory. This led to Einstein winning the Nobel Prize in Physics in 1921 "for his discovery of the law of the photoelectric effect."[2]

Stage 2: developing the silicon PV cell

In the early 1950s, Daryl Chapin, Calvin S. Fuller, and Gerald Pearson at Bell Laboratories developed the first modern solar PV cell using silicon (with 4% efficiency; they later achieved 11%).[3] On April 25, 1954 in Murray

Hill, New Jersey, they demonstrated their silicon solar module by using it to power a toy Ferris wheel and a radio transmitter. The first practical use of the technology was to power satellites in space, starting in 1958 with the Vanguard I, which used a small array (less than one watt) for its radios.[4]

Stage 3: low-cost manufacturing

As shown in Figure 1.1, the unit price of a solar cell fell over time. In the 1970s, Dr. Elliot Berman, with help from Exxon Corporation, designed a production process that sharply reduced costs, slashing the price from $100 per watt to around $20 per watt. New practical uses were found for the silicon PV cell, such as powering remote sensors and navigation lights on offshore oil rigs. In 1973, the University of Delaware showed that solar energy could power a house. In 1978, NASA installed the first community PV system, which supplied electricity to 15 homes in the isolated Papago Indian Reservation in Arizona for water pumping and residential needs.

Solar energy took another major step forward in the early 1980s when the oil company Arco Solar built the first large-scale solar park in Hesperia, California, near Los Angeles. Opened in 1982, this solar park generated 1 megawatt, or 1,000 kilowatts, supplying the Pacific Gas & Electric (PG&E) Company with enough power for 2,000–2,500 homes. Until this point, the industry had only achieved 15% efficiency in harnessing energy from sunlight. However, this changed in 1992 when the University of South Florida unveiled a 15.9% efficient thin-film PV cell made of cadmium telluride.[5] In the 1990s, as cell efficiencies rose and manufacturing costs declined, utilities began to take interest in PV systems. PG&E became the first U.S. utility to operate a large-scale PV system, a 500 kW array in Kerman, California. The early 2000s saw a rise in grid-tie systems for residential and commercial customers, and, as shown in Figure 1.2, by 2012 solar reached grid parity with conventional fuels.

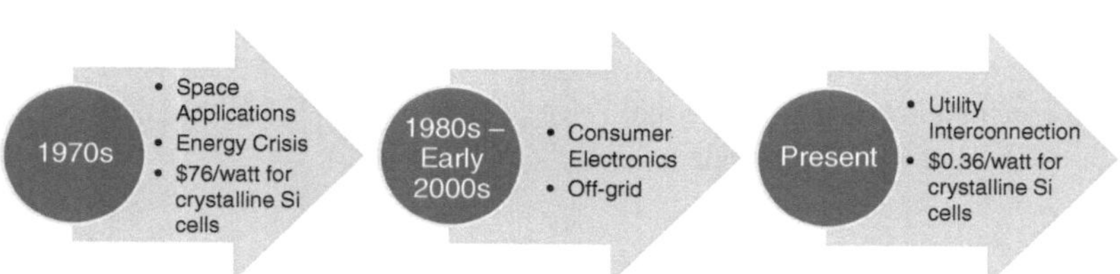

Figure 1.1 The progression of manufacturing of solar PV cells from the 1970s to today.

Source: Author.

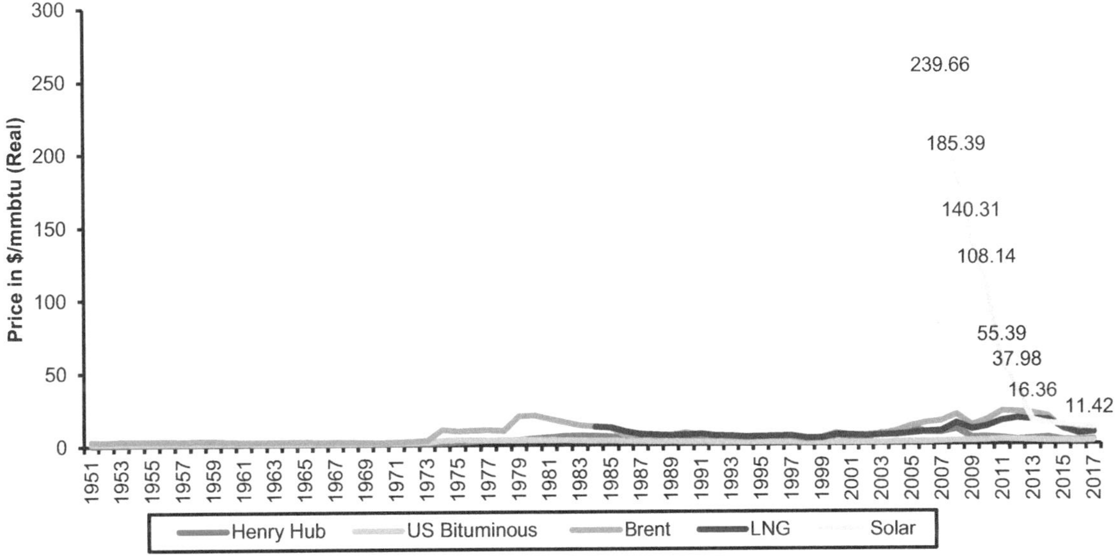

Figure 1.2 Achieving grid parity – solar PV's rapid cost decline beginning in 2009 and current cost equivalence with conventional sources (price in $/mmbtu).

Source: Michael Parker and Flora Change, Bernstein. Data from EIA, CIA, World Bank.

By 2016, solar cells could be manufactured as thin as paper using an industrial printer and with 20% efficiency, and a single strip could produce up to 50 watts per square meter.[6] Figure 1.3 shows the dramatic cost reduction achieved in PV materials and manufacturing, yet soft costs remain high.

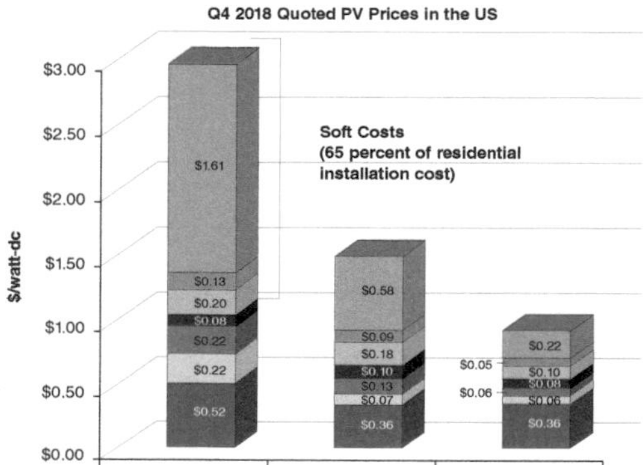

	Residential	Commercial	Utility-Scale
Supply Chain, Overhead, Margin	$1.61	$0.58	$0.22
Engineering and PII	$0.13	$0.09	$0.05
Direct Labor	$0.20	$0.18	$0.10
Structural BOS	$0.08	$0.10	$0.08
Electrical BOS	$0.22	$0.13	$0.06
Inverter	$0.22	$0.07	$0.06
PV Module	$0.52	$0.36	$0.36

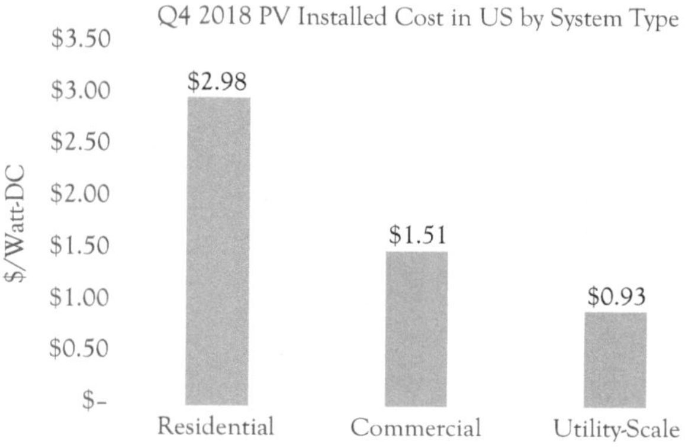

Figure 1.3 Q4 2018 installed PV prices in the U.S. by system size. The figure on the top shows that soft costs comprise 65% of residential installation cost. The figure on the bottom shows that economies of scale enable lower prices for commercial and utility-scale systems.

Source: *Harness It: Renewable Energy Technologies and Project Development Models Transforming the Grid* by Michael Ginsberg.

Data from Wood Mackenzie, Limited/SEIA U.S. Solar Market Insight Q4/2018®.

As of Q1 2018, the U.S. has 55.9 GW of total installed capacity, with 18% of all electricity in the U.S. coming from renewable sources and most new capacity from solar.[7]

New Generation In-Service (New Build and Expansion)

Primary Fuel Type	December 2018		January–December 2018 Cumulative		January–December 2017 Cumulative	
	No. of Units	Installed Capacity (MW)	No. of Units	Installed Capacity (MW)	No. of Units	Installed Capacity (MW)
Coal	0	0	4	10	3	45
Natural Gas	7	3,134	103	20,048	106	12,531
Nuclear	0	0	5	350	1	102
Oil	0	0	14	25	37	89
Water	0	0	10	33	14	221
Wind	12	1,943	55	6,028	83	7,415
Biomass	0	0	13	68	27	272
Geothermal Steam	2	62	4	82	2	55
Solar	15	502	429	4,181	750	6,186
Waste Heat	0	0	2	51	1	220
Other *	7	0	33	5	28	1
Total	43	5,641	672	30,881	1,052	27,137

"Other" includes purchased steam, tires, and miscellaneous technology such as batteries, fuel cells, energy storage, and fly wheel.

Figure 1.4 Table showing the number of units and installed capacity of different energy sources in the U.S. installed in Q1 2018 vs Q1 2017. Solar was the third leading source of new capacity in the U.S. in 2018.

Source: Office of Energy Projects Energy Infrastructure Update for December 2018, Federal Energy Regulatory Commission.[8,9]

The importance of O+M and M+V

The rapid deployment of solar energy, both PV and concentrating solar power (CSP), has led to a correspondingly swift increase in employment in the sector. The number of employees who spent a portion of their time working on solar energy rose from 93,000 in 2010 to almost 350,000 in 2017 – about 250,000 (or 70%) of whom spent at least half of their time supporting the solar portion of their business.[10] This rate of growth is more than nine times faster than employment growth in the overall U.S. economy.[11]

In 2017, solar energy workers represented about 40% of the total electric power generation workforce, which is more than the combined number of workers with fossil fuels (gas, coal, and oil) in electricity generation.[12] While the heaviest concentrations of jobs in the U.S. are in California, Massachusetts, New York, Texas, Florida, and Arizona, employment is expanding rapidly throughout the U.S.[13]

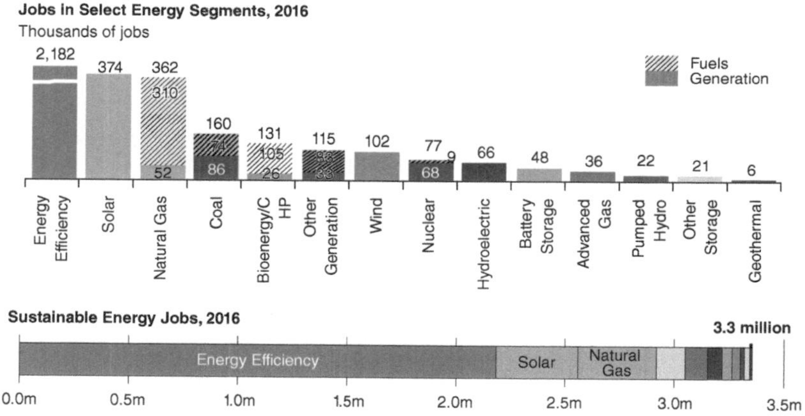

- The renewable, energy efficiency, and natural gas sectors employed an estimated 3.3 million Americans in 2016, according to the Department of Energy. Energy efficiency alone supported 2.2 million jobs, while solar supported roughly 374,000 and natural gas 362,000.
- While renewable sectors like solar, wind, hydropower, and geothermal do not require upstream processing or extraction of a fuel, fossil-fired generation does. Adding in these fuel-related jobs notably boosts the total employment by fossil fuel-fired generation and bioenergy. In 2016, 86% of the 362,000 jobs associated with the natural gas sector came from fuel supply. Coal employed 160,000, with about half in coal production and supply.
- Energy efficiency jobs related to construction often hire people who also work on other types of construction tasks (26% of the 1.4 million employees in this category spend only the minority of their time on efficiency).

Figure 1.5 Solar jobs in the U.S. were the second highest of all energy segments in the U.S. in 2016.

Source: Bloomberg New Energy Finance.

This trend is not limited to the U.S. Solar is emerging globally as one of the leading new sources of jobs. The International Renewable Energy Agency (IRENA) estimates the solar PV industry accounts for nearly 3.4 million jobs worldwide.[14]

Unlike fuel generators that require weekly or even daily maintenance, once installed, silent PV systems can be easy to forget. Aside from trackers, they are comprised of steady-state devices with no moving parts. However, PV O+M and M+V is becoming increasingly important as the industry grows. There is a rising need not only for engineers and installers, but also for maintenance personnel who know how to diagnose, operate, and maintain PV systems. Minor errors can severely impact a project's bottom line.

This book distills the best practices in PV inspection, O+M and M+V. In addition, it outlines the future of O+M practices and technology in the industry.

Case study: verifying system size to avoid mistakes that erode the bottom line

In a country with ample sunlight and where electricity costs $0.40+ per kWh, solar energy is a "no brainer." Recognizing these attractive economics, a facility in Curaçao, a Caribbean island in the former Kingdom of the Netherlands, installed a PV array to offset, with energy efficiency upgrades, 100% of consumption to become *net zero*, producing enough to cover all electrical consumption. In designing the system, the designers used three years of billing (consumption) history as reported by Aqualectra, the local utility. Unfortunately, the kWhs reported by the utility were under half of the actual consumption – the utility had been reading the meter incorrectly and undercharging for years. Aqualectra limited the PV production to 82% of the load. The system was sized based on these erroneous numbers, and as a result now meets less than half of the facility's energy needs. Figure 1.6 shows solar production does not come close to demand.

To add insult to injury, the utility then claimed back fees on the past three years of undercharged bills *and* slapped a monthly standby tariff on the PV system of $17.92 per kW per month, further diminishing the return on investment.

The 125 kW system cost about $437,500 to install. Considering the high upfront cost, an independent energy audit by the stakeholders would have been prudent. Connecting current transformers (CTs) to the facility's circuits for a period of one to three months before confirming the system sizing calculations would have revealed the discrepancy between the utility's bills and actual consumption. In general, facility-wide energy audits

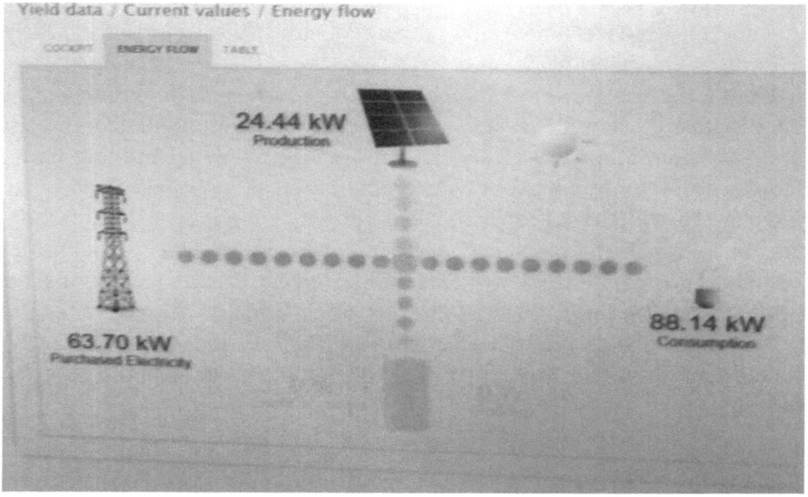

Figure 1.6 PV yield data showing only 28% of load met by PV production.
Source: Author, taken from Solar-Log dashboard.

are advisable when spending over $100,000 on system upgrades. The following pages show in detail the difference in **net present value (NPV)** between the design, actual, and redesigned systems.

1) DESIGN: Cost-benefit analysis

Net Present Value (NPV) = Present Value (PV) of Benefits – PV of Costs
= $464,428.81 – $437,500 – $1,562.21
= **$25,366.60**

Assumptions for PV calculations:

- 4% discount rate (i)
- 25-year period (n)
- Annual compounding

BENEFITS

Annual Benefits = $29,729
PV of Benefits = $464,428.81

COSTS

PV of Installation Cost = $437,500 (based on $3.50 per watt, $3.50 × 125,000 watts)
O+M & M+V = $100 per month

PV of O+M & M+V = $1,562.21

Annual Benefits = Annual Cost of Electricity without Solar – Annual Cost of Electricity with Solar
= \$90,587 – \$60,858
= \$29,729

Annual Cost of Electricity without Solar = Annual Facility Consumption × \$ per kWh
= 212,546 kWh (109,960 kWh + 102,586 kWh) × \$0.4262/kWh average cost per kWh
= \$90,587

Annual Cost of Electricity with Solar = Annual Costs – Annual Benefits
= \$73,745 – \$12,887
= \$60,858

Annual Electricity Cost with Solar = \$46,865 + \$26,880
= \$73,745

Annual Costs of Electricity with Solar Breakdown Estimated Cost During Non-Sunlight Hours[1]
= 109,960 kWh × \$0.4262/kWh average cost per kWh
= \$46,865

Aqualectra's "Stand-by Tariff" = Array Size × \$17.92/kW/Month × 12 Months
= 125 kW × \$17.92/kW/month × 12 months
= \$26,880

Annual benefits

Estimated Benefit from Solar = Estimated Annual Energy Production of 125 kW Array – Estimated Energy Use During Sunlight hours × \$0.18 Feed-in Tariff

Energy Production from 125 kW Array During Sunlight Hours
= **174,180 kWh**

This estimate comes from the formula:

Estimated Production (kWh) = PV kW size × 365 days × 4.5 peak sun hours (PSH) × 0.85
= 125 kW × 365 days per year × 4.49 PSH × 0.85
= **174,180 kWh**

Estimated Energy Use During Sunlight Hours = **102,586 kWh**
Amount Exported = **174,180 kWh – 102,586 kWh**
= **71,594 kWh**

Value of Exported Solar = **Amount Exported** × **Feed-In Tariff (Amount Compensated by Aqualectra)**
= **71,594 kWh** × **\$0.18**
= **\$12,887**

Now, we know that the actual demand was about two times the design demand. Let's see what this means for our investment.

2) ACTUAL: Cost-benefit analysis

Based on *actual* energy consumption (double design, same time value of money assumptions:

Net Present Value (NPV) = Present Value (PV) of Benefits – PV of Costs
= \$739,783.60 – \$437,500 – \$1,562.21
= **\$300,721.39**

HOWEVER, we now are spending \$133,819 per year with solar whereas in the design we were spending \$60,858 per year with solar. That's a difference of \$72,961. The present value of the extra amount that will now need to be spent that could have been offset is \$1,139,802.57. Subtract that from our NPV:

\$300,721.39 – **\$1,139,802.57**
= **–\$839,081.18 (negative present value)**

Annual Savings = Annual Cost of Electricity without Solar – Annual Cost of Electricity with Solar
= \$181,174 – \$133,819
= **\$47,355 (PV = \$739,783.60)**

Annual Cost of Electricity without Solar = Annual Facility Consumption × \$ per kWh
= 425,092 kWh (212,546 kWh × 2 (actual demand is about double design) × \$0.4262/kWh average cost per kWh)
= \$181,174

Annual Cost of Electricity with Solar = Annual Costs – Annual Benefits

= $133,819 – $0 (note costs go from $73,745 to $133,819 as far less consumption is offset and no PV is exported for revenue)

= $133,819

Annual Facility Electricity Cost with Solar = $93,730 + $26,880 + $13,209

= $133,819

Annual Costs of Electricity with Solar Breakdown Estimated Cost During Non-Sunlight Hours[1]

= 109,960 kWh × 2 × $0.4262/kWh average cost per kWh

= $93,730

Aqualectra's "Stand-by Tariff" = Array Size × $17.92/kW/Month × 12 Months

= 125 kW × $17.92/kW/month × 12 months

= $26,880

Estimated Energy Use During Sunlight Hours = 205,172 kWh (102,586 kWh × 2)

Estimated Production = 174,180 kWh

205,172 – 174,180

= 30,992 kWh **still needed** × $0.4262/kWh

= $13,209

If we had designed the system to meet actual demand and made it 250 kW:

3) REDESIGN: Cost-benefit analysis

NPV = PV of Benefits – PV of Costs

We would have a NPV = $928,565.50 – $878,124.42

= **$50,441.08**

This is double the design case NPV, although we will still have a difference of $60,877 between what we are now spending per year with solar and what we were spending in the design case ($121,735–$60,858), which is a PV of $951,025. Subtract that from our NPV:

= $50,441.08 – $951,025

= **–$900,584 (negative present value)**

Costs

= $875,000 (upfront, double design although maybe less because of economies of scale) + $3,124.42 (PV of $200 per month for O+M & M+V, double before)

= $878,124.4

Annual Savings = Annual Cost of Electricity without Solar – Annual Cost of Electricity with Solar

$= \$181{,}174.21 - \$121{,}734.85$

$= \$59{,}439.36$

PV of Annual Savings = $928,565.50

Electricity Cost without Solar

$= 212{,}546 \text{ kWh} \times 2 \times 0.4262/\text{kWh}$ average cost per kWh

$= \$181{,}174.21$

Annual Cost of Electricity with Solar = Annual Costs – Annual Benefits

$= \$147{,}489.90 - 25{,}755.05$

$= \$121{,}734.85$

Annual Facility Electricity Cost with Solar

$= \$93{,}729.90 + \$53{,}760$

$\boxed{= \$147{,}489.90}$

Annual Costs of Electricity with PV Breakdown Estimated Cost During Non – Sunlight Hours[1]

= 109,960 kWh × 2 × $0.4262/kWh average cost per kWh

= $93, 729.90

Aqualectra's "Stand by Tariff" = Array Size × $17.92/kW/Month × 12 Months

= 250 kW × $17.92/kW/month × 12 months

= $53,760

Benefits

Annual benefits = $25,755.05 (modest increase is due to feed-in tariff being 43% of customer rate)

Estimated Energy Use During Sunlight Hours = **205,172 kWh (102, 586 kWh * 2)**

Estimated Solar Production = **348, 255.63 kWh** (250 kW × 4.49 PSH × 365 days × 0.85)

Amount Exported = **348, 255.63 kWh – 205,172 kWh = 143, 083.63 kWh**

Value of Exported PV = **Amount Exported × Feed – In Tariff (Amount Compensated by Aqualectra)**

= 143, 083.63 kWh × $0.18

= $25, 755.05

There are many lessons to be learned from this case study. Clearly, the standby tariff per kW per month, low feed-in tariff compared to the cost of electricity on the island, coupled with the fact that the system has no battery storage and cannot use the energy to reduce consumption during non-daylight hours, calls into question the economics of this installation, unless it is a large solar plant (e.g. 1 MW). However, even if the system could be increased to a larger size, which it cannot due to utility restrictions, the standby fee would essentially offset the gain from the additional energy exported.

It is important for stakeholders to carefully understand not only the energy consumption of their facility but also the economics of how solar is valued by the utility and of how battery storage could have a significant economic benefit, despite its upfront cost.

The NABCEP PV System Inspector credential

The North American Board of Certified Energy Practitioners (NABCEP) is a leading certification organization for renewable energy professionals. As evidence of the need for improved standards in PV **measurement and verification (M+V)**, the organization launched the PV System Inspector (PVSI) credential in May 2017 to raise standards for professional PV inspectors and representatives of **Authorities Having Jurisdiction (AHJs)**.

PVSIs are expected to be knowledgeable about a PV system's components and their operation, codes and ordinances, and safety. Inspectors primarily work in the private sector or for local permitting offices. These individuals provide quality control by analyzing and interpreting design plans and building documents, conducting on-site inspections, and reporting results.

Individuals seeking to demonstrate their competence as PVSIs can now take the exam on the NABCEP website. More information on this is provided in Chapter 5, including sample exam questions.

Notes

1 U.S. Department of Energy. 2019. "The History of Solar." *www1.Eere.Energy.Gov*. https://www1.eere.energy.gov/solar/pdfs/solar_timeline.pdf.
2 "The Nobel Prize in Physics 1921." 2019. *Nobelprize.Org*. www.nobelprize.org/prizes/physics/1921/summary/.
3 "This Month in Physics History." 2019. *Aps.Org*. www.aps.org/publications/apsnews/200904/physicshistory.cfm.
4 U.S. Department of Energy. 2019. "The History of Solar." *www1.Eere.Energy.Gov*.
5 Ibid.
6 Baker, A. 2019. "History of Solar Cells: How PV Panels Evolved | Solar Power Authority." *Solar Power Authority*. www.solarpowerauthority.com/a-history-of-solar-cells/.

7 "Renewable Sources Account for Most New U.S. Power Capacity." 2019. *Forbes. Com*. www.forbes.com/sites/rrapier/2018/05/06/renewable-sources-account-for-most-new-u-s-power-capacity/#2f9510a25971.

8 Office of Energy Projects Energy Infrastructure Update for December 2018 FERC. www.ferc.gov/legal/staff-reports/2018/dec-energy-infrastructure.pdf.

9 "Renewable Sources Account for Most New U.S. Power Capacity." 2019. *Forbes. Com*.

10 *2018 U.S. Energy and Employment Report. Report*. NASEO and Energy Futures Initiative. 2018. Accessed March 25, 2019. www.usenergyjobs.org/report. p. 47.

11 *2017 National Solar Jobs Census*. Report. The Solar Foundation. November 2017. Accessed March 25, 2019. www.thesolarfoundation.org/national/. pp. 6, 52.

12 *2018 U.S. Energy and Employment Report*. NASEO and Energy Futures Initiative. pp. 38–39.

13 *2017 National Solar Jobs Census*. Report. Solar Foundation. 2017.

14 "Renewable Energy Jobs Reach 10.3 Million Worldwide in 2017." IRENA" International Renewable Energy Agency. Accessed March 25, 2019. https://irena.org/newsroom/pressreleases/2018/May/Renewable-Energy-Jobs-Reach-10-Million-Worldwide-in-2017.

2 Electricity and solar cell and system fundamentals

Before diagnosing systems, performing maintenance, or implementing corrective actions to resolve issues, it is important to understand the fundamentals of electricity and solar photovoltaics. This chapter provides a quick overview of the basics of electricity, how solar energy is converted from sunlight to electricity, and the components of solar energy systems.

Electricity basics

The components of electricity

What exactly *is* **electricity**? It is often referred to as the "invisible force," which makes it seem magical. Unlike water flowing through a pipe, electricity is difficult to imagine: we cannot see or hear it, and we know better than to try and touch it! It is only in the past 200 years or so of human history that we have been able to grasp and manipulate the forces that give rise to electricity.

The commercialization of electricity by Thomas Edison and Nicola Tesla happened even more recently, about 100 years ago, which is nearly the blink of an eye in the timescale of our existence. While the technology may seem commonplace today, imagine showing a light to someone from the Middle Ages- you might be hanged for witchcraft!

It is important to retain this perspective and wonderment while learning the science of electricity. Our knowledge today rests on centuries of inquiry.

Remember the building block of matter, the atom? It is comprised of three types of sub-atomic particles: positively-charged protons, neutral neutrons, and negatively-charged electrons. The protons and neutrons are located in a central body called the nucleus, and the electrons move around it in orbits, as shown in Figure 2.1.

Since opposites attract, negatively-charged electrons are held close to the nucleus by the force of their attraction to the positively-charged protons.

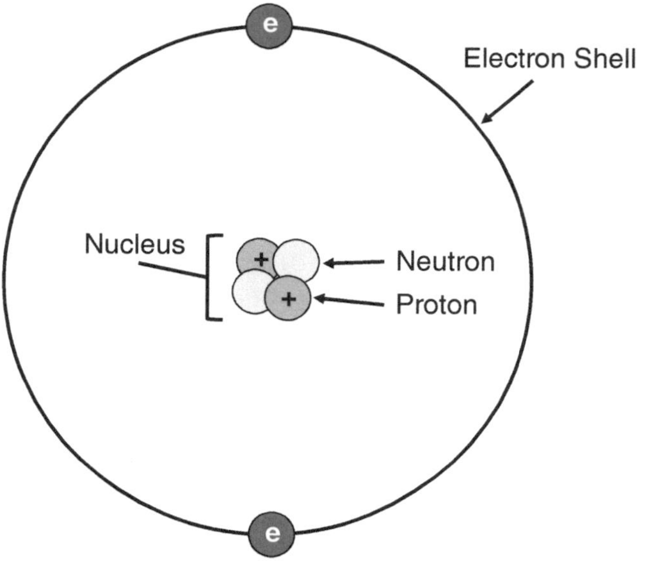

Figure 2.1 Atomic model of a helium atom, showing positions and charges of neutrons, protons, and electrons.

Source: *Harness It: Renewable Energy Technologies and Project Development Models Transforming the Grid* by Michael Ginsberg.[1]

In 1800, Italian Physicist Alessandro Volta discovered that if you introduce an external charge to an atom you can "eject" stable electrons out of their orbits and use them in a circuit. This discovery led to our definition of electricity as the free flow of electrons through a conductor, or wire. We put those electrons to work when they power our **loads**, such as lights and motors.

Although an imperfect analogy, electron flow can be likened to water flowing through a pipe. Electrons flow due to "pump pressure" provided by a **voltage source**, through a "pipe," or **conductor**. Along the way those electrons encounter **resistance**, which makes it harder for them to flow, like obstructions within the pipe, and the electrons are used to power our loads, not unlike how machinery in a watermill is turned by harnessing the power of running water. Figure 2.2 shows a closed circuit in which electrons are conducted by wires from the voltage source (battery) to the electrical load (light) and then back to the voltage source again, encountering resistance along the way.

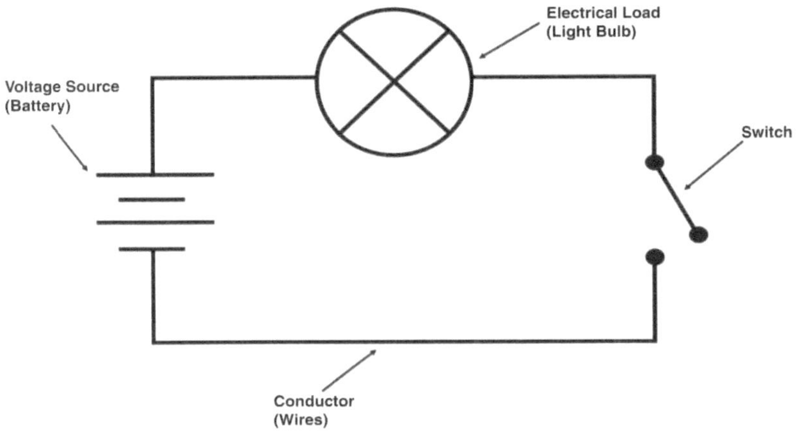

Figure 2.2 Simple electrical circuit showing a voltage source (battery), conductor (wires), and an electrical load (light bulb).

Source: *Harness It: Renewable Energy Technologies and Project Development Models Transforming the Grid* by Michael Ginsberg.

Power versus energy

It is important to understand the difference between power and energy. While these terms are often used interchangeably, they are in fact distinct. When we talk about **power** we are referring to the capacity of a generator in watts. When purchasing a lamp, for instance, 60 watts refers to the amount of energy it pulls per unit time.

On the other hand, **energy** is power multiplied by time, and its unit is watt-hours or kilowatt hours (kWh). For instance, if your 60-watt lamp is used for 4 hours per day, the lamp uses 60 watts over 4 hours:

$$\text{Power (watts)} \times \text{Time (hours)} = \text{Energy (kWh)}$$
$$60 \text{ watts} \times 4 \text{ hours} = 240 \text{ watt hours (Wh) per day}$$
$$240 \text{ watt hours} \div 1000 = 0.24 \text{ kilowatt hours (kWh) per day would be the electrical demand from your lamp}$$

But what about the other way around? If you have a solar energy system on your roof with a power capacity of 500 watts and it works for 6 hours per day, then how much energy will be *produced*?

$$\text{Power (watts)} \times \text{Time (hours)} = \text{Energy (kWh)}$$
$$500 \text{ watts} \times 6 \text{ hours} = 3000 \text{ watt hours (Wh) per day}$$
$$3000 \text{ watt hours (Wh)} \div 1000 = 3 \text{ kilowatt hours (kWh) per day}$$

Don't forget, you can also rearrange the equation:

$$\text{Power (watts)} = \text{Energy(kWh)} \div \text{Time (hours)}$$

Basic circuit analysis: series versus parallel circuits

The components of electricity and their relationship to one another were discovered by the 18th-century German physicist George Ohm. These components are voltage (V) in volts, current (I) in amps and resistance (R) in ohms.

The formula is **V = IR.**

To visualize this relationship, let's go back to our water pipe analogy, as shown in Figure 2.3. **Voltage**, the electrical potential, is the product of **current** and resistance. The electrical potential of a circuit is based on its current (the flow of electrons), and its resistance (how hard it is for the electrons to flow). Resistance represents electrical potential "lost" in the form of heat.

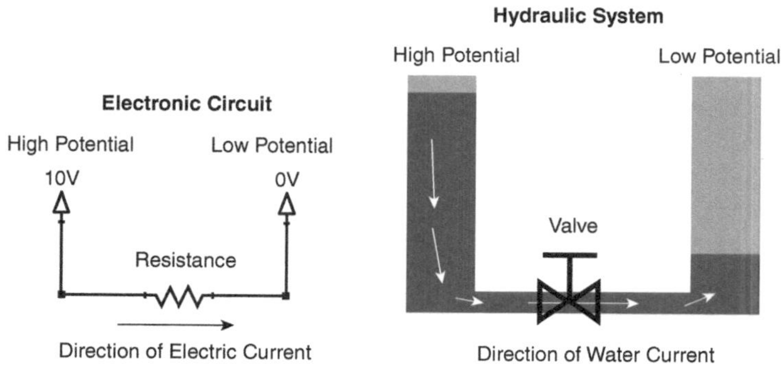

Figure 2.3 Comparing hydraulic flow to electrical flow.
Source: ikalogic.

The **Ohm's Law** triangle in Figure 2.4 is very useful when calculating this relationship. To solve for the missing variable cover it up with your finger.

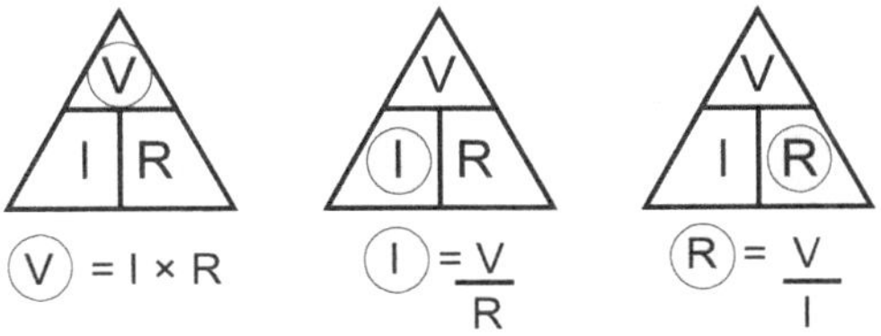

Figure 2.4 Ohm's Law triangle. To solve for the missing variable cover it up with your finger.

Source: *Harness It: Renewable Energy Technologies and Project Development Models Transforming the Grid* by Michael Ginsberg.

Let's look at a couple of examples using *series* and *parallel* circuits.

Series – in a series circuit the current flows along one path. In Figure 2.5, the resistors, which may be lamps, are placed *in series*, one after the other. In a series circuit, the total resistance is the sum of the individual resistances. For instance, let's say you have a series circuit where the voltage is 12v, resistor 1 (R_1) is 50 ohms and resistor 2 (R_2) is 150 ohms. To find the total resistance (R_T) you add $R_1 + R_2 = 200$ ohms. To find the current, using Ohm's Law, $I = V/R$, thus $I = 12v/200$ ohms $= 0.06$ amps $\times 1000 = 60$ milliamps (mA). In a series circuit, the total voltage is the sum of the individual voltages. An important concept in a series circuit is that if there is a fault with one load, the entire circuit is broken (opened) and all subsequent loads will not be powered (remember going through your Christmas lights trying to find the one that was broken? That's a series circuit!).

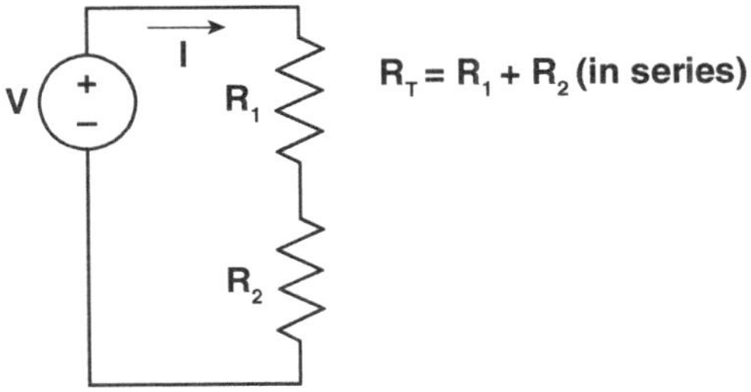

Figure 2.5 A series circuit showing two resistors connected to a voltage source.

Source: *Harness It: Renewable Energy Technologies and Project Development Models Transforming the Grid* by Michael Ginsberg.

Parallel – on the other hand, a circuit that has multiple branches divides current across the different branches at the same time, so if one branch has a fault, current can still flow. In a parallel circuit, like the one shown in Figure 2.6, the total resistance is the sum of the reciprocal of each resistor.

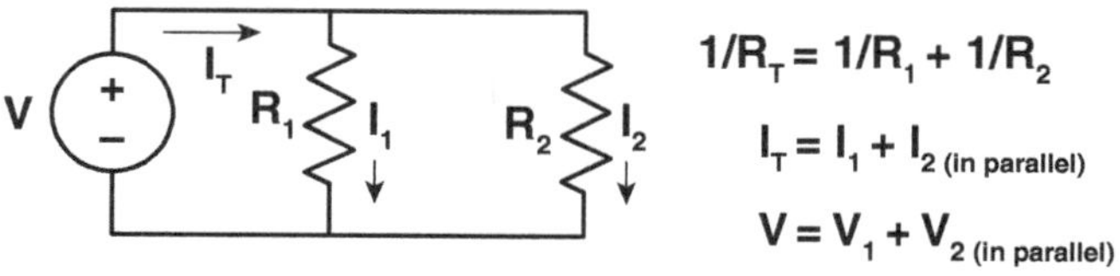

Figure 2.6 A parallel circuit showing two resistors connected to a voltage source.

Source: *Harness It: Renewable Energy Technologies and Project Development Models Transforming the Grid* by Michael Ginsberg.

This means that the total resistance in a parallel circuit is always less than the smallest individual resistance. This makes sense because more paths for electrons to flow means greater current. Greater current means less resistance, and indeed total current in a parallel circuit is the sum of the individual branch currents.

Let's take an example. If R_1 is 2 ohms and R_2 is 3 oms, then the R_T is 1/2 ohms + 1/3 ohms. Finding the common denominator, $1/R_T = 3/6$ ohms + 2/6 ohms = 5/6 ohms. Taking the reciprocal of $1/R_T$, the total resistance is thus 6/5 ohms = 1.2 ohms. Finding the total current, we would use Ohm's Law, I = V/R, and thus if the voltage = 6v, then I = 6v/1.2 ohms = 5 amps.

If you know the branch currents and take into account the fact that voltage is the same across branches, an easier way to find total resistance is just to use the resistance of each branch to find the individual branch current and then the total current.

$$\mathbf{V = IR \rightarrow I_T = V / R_T \rightarrow V / R_1 + V / R_2}$$
$$\mathbf{I_1 = V / R_1 = 6V / 2\ \Omega = 3amp}$$
$$\mathbf{I_2 = V / R_2 = 6V / 3\ \Omega = 2amp}$$
$$\mathbf{I_T = I_1 + I_2 = 3amp + 2amp = 5\ amp}$$

Another important concept is the Power Triangle, shown in Figure 2.7, which shows that power equals the product of voltage and current.

The Ohm's Law wheel in Figure 2.8 shows all of the relationships between power, voltage, current, and resistance. The inner circle contains the missing variables, which can be solved for using two of the other variables.

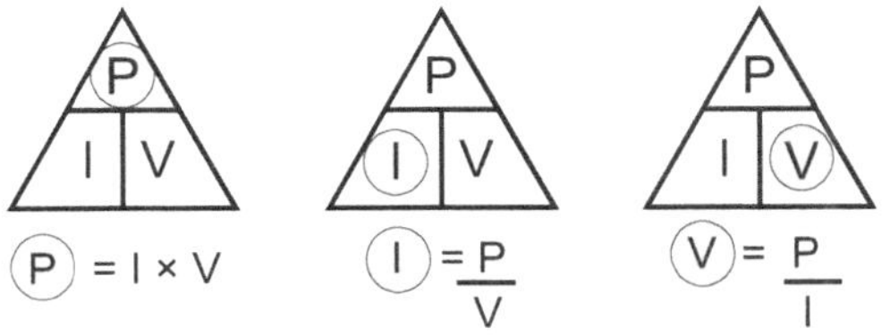

Figure 2.7 The Power Triangle. To solve for the missing variable cover it up with your finger.

Source: *Harness It: Renewable Energy Technologies and Project Development Models Transforming the Grid* by Michael Ginsberg.

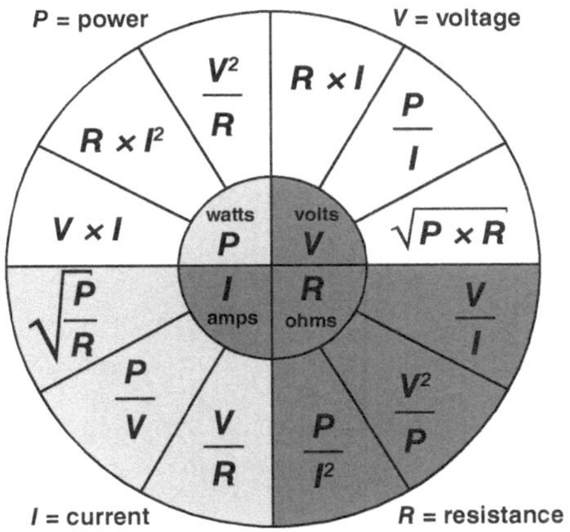

Figure 2.8 Ohm's Law wheel. To solve for the missing variable cover it up with your finger.

Source: Wikimedia Commons.[2]

Railway Method (dimensional analysis) for unit conversion

When calculating elements of electricity one useful way to keep track of the units is called dimensional analysis, or the Railway Method. It is so called because we place the units we are multiplying or dividing across like train tracks. Always write out the units and treat them as factors that can be canceled.

For instance, if you want to convert 30 feet into inches:

Conversion factor: 1 ft = 12 in

30 ~~ft~~	12 inches
	1 ~~ft~~

The feet cancel and you are left multiplying 30 × 12 = 360 inches.

Now, should there be more or less inches than the equivalent in feet? More! Because there are 12 inches in 1 foot. Also, we should be left with the unit inches.

In metric units this is much easier. Let's say you want to convert 10 meters (m) to centimeters (cm).

Conversion factor: 1 meter = 100 centimeters

10 ~~m~~	100 cm
	1 ~~m~~

The meters cancel and you are left multiplying 10 × 100 = 1000 cm.

When dealing with multiple unit conversions this method is very helpful. For instance, if we want to convert 6.5 yards to millimeters (mm):

Conversion factors:

1 yard = 3 feet
1 foot = 12 inches
1 inch = 25.4 mm

6.5 ~~yards~~	3 ~~feet~~	12 ~~inches~~	25.44 mm
	1 ~~yard~~	1 ~~foot~~	1 ~~inch~~

This results in the calculation 6.5 × 3 × 12 × 25.44 = 5,952.96 mm.

Common sense check: There are 3 feet in a yard, 12 inches in a foot, and many millimeters in an inch, so we should end up with a much larger number. The yards, feet, and inches should cancel out, and we should be left with millimeters.

Direct current (DC) versus alternating current (AC)

Electricity comes in two forms – **direct current (DC)** and **alternating current (AC)**. Originally, Thomas Edison developed DC, in which, as the name implies, electrons travel *directly* from the voltage source (in Edison's case a battery) to the load. The same electrons generated from the source are used by the load. Note that batteries and solar cells produce DC.

Edison's protégé-turned-rival, Nicola Tesla, developed another form of electricity, which allowed electricity to be generated far from the loads. This model allowed for mass electrification of the U.S. and Europe throughout the mid-20th century. In AC, electrons "jiggle" back and forth in a sinusoidal (sine) wave, and the electrons at the source are not the ones used by the load.

Today, with the proliferation of rooftop solar and other distributed energy sources, and the advancement of power electronics, DC is becoming an alternative primary source of electricity. For now, however, **inverters** are required with all solar PV systems that export power to the grid or power appliances that use AC. The primary role of these inverters is to convert the DC electricity to AC.

Solar energy basics

Measurement and production

Solar irradiation can be considered as fuel. Due to night, clouds, and other obstructions, during a 24-hour day we receive a limited number of hours of "peak sun," defined as 1,000 watts/m^2. Globally, this ranges from two peak sun hours at the poles to eight peak sun hours close to the equator.

The amount of horizontal irradiation that hits a particular area can be measured by a pyranometer, as shown in Figure 2.9, and varies according to where it is on the globe, as shown in Figure 2.10. You can therefore

Figure 2.9 Pyranometer. Used to measure the solar irradiation at a site in watts per m^2.

Source: Hukseflux Thermal Sensors. https://commons.wikimedia.org/wiki/File:SR20_pyranometer_1.jpg.[3]

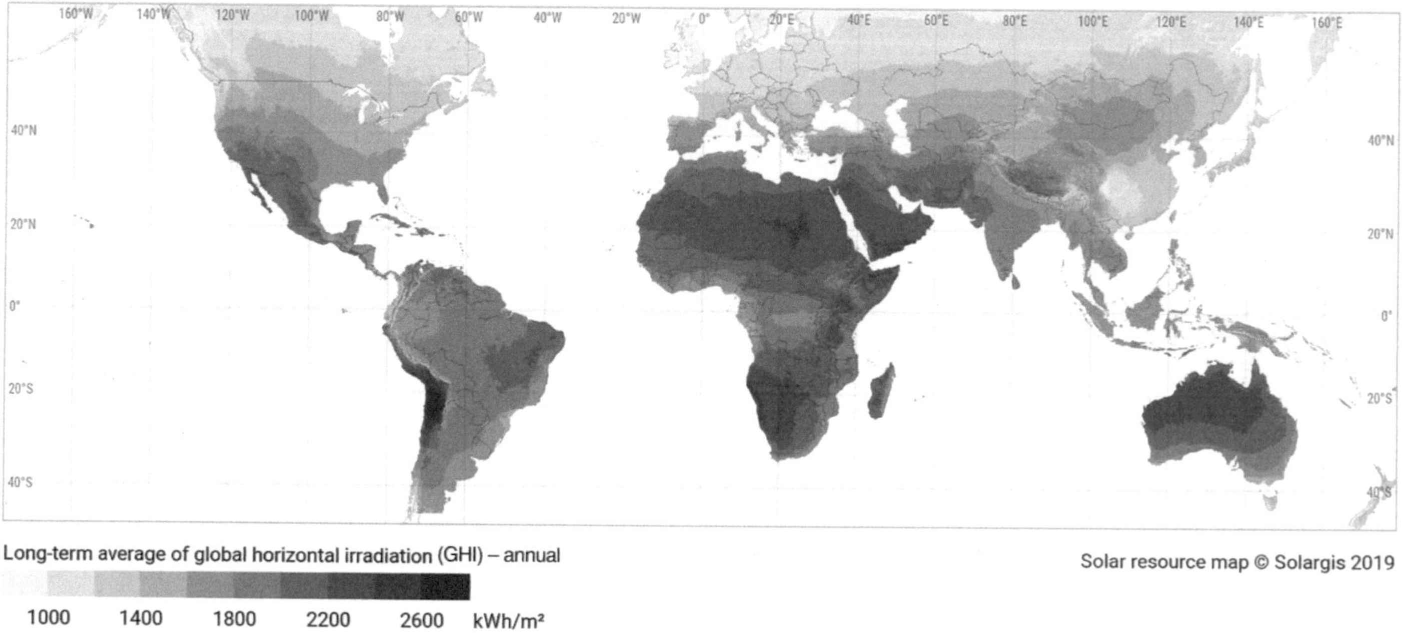

Figure 2.10 Map showing global horizontal irradiation. Divide GHI by 365 for daily peak sun hours.

Source: SolarGIS.[4]

estimate the anticipated production of a PV system production based on your location, using the equation:

$$\textbf{Annual PV System Production (kWh)} = \text{PV system size (kW)} \times \text{number of peak sun hours (PSH)}^{*} \times 365 \text{ days}$$

You need to then derate for losses throughout the system (a 15% loss is typically used, so we multiply by 85% since that is the amount retained).

$$\textbf{Annual PV System Production (kWh)} = \text{PV system size (kW)} \times \text{number of peak sun hours (PSH)} \times 365 \text{ days} \times 0.85 \text{ (15\% derating)}$$

Determining the power rating in kW of a grid-tie PV system is straightforward. We start with demand, based on two years of energy consumption (kWh). For example, the average annual energy consumption of an American household is 6,515 kWh. Therefore, average household annual PV system production size can be calculated based on the need to meet 6,515 kWh of demand.

The kWh produced should equal demand. Given the PV system location and annual household or facility energy consumption, the only missing variable is system size:

$$\textbf{PV kW System Size} = \text{Annual PV System Production}\left(\text{kWh}\right) + \left(\text{PSH} \times 365 \text{ days} \times 0.85\right)$$

So, for a location with 6 PSH needing to produce 6,515 kWh, the system size, or capacity, would be 3.5 kW.

$$6{,}515 \text{ kWh} \div \left(6 \text{ PSH} \times 365 \text{ days} \times 0.85\right) = 3.5 \text{ kW}$$

The importance of latitude and azimuth

Latitude is a critical factor in determining the amount of sun available to a site. At higher latitudes the sunlight is distributed over greater area, while closer to the equator sunlight is more concentrated. This is why the PSH is higher at the equator than at the poles. It also informs the tilt angle of the array. At the equator, the sun is directly overhead and thus the array should be zero tilt, or flat. At north and south latitudes, the tilt angle is equal to the latitude of the site, called *latitude tilt*. This means that when we venture away from the equator, the modules are tilted towards the equator for optimal annual production. For a rule of thumb, in the summer, the sun is high

* Peak sun hours is the daily average of global horizontal irradiation. Figure 2.10 gives the annual average ranging from 1,000 to 2,600 kWh per m^2. Thus, the daily peak sun hours are this number divided by 365, which ranges from 2.7 to 7.1.

in the sky and the array is tilted at latitude minus 15 degrees. In the winter the sun is low and the array is tilted at latitude plus 15 degrees.

The *solar azimuth angle*, or direction from which sunlight arrives at the array, is another important determinant of a system design. In the Northern Hemisphere, the sun primarily passes through the sky from the south, and in the Southern Hemisphere, from the north.

Box 2.1 How the voltage and current of a solar module is determined

- *Voltage* of a solar cell is sensitive to temperature and based on the material's energy bandgap
  - A material's energy bandgap determines its voltage (more on this below). The lower the temperature the higher the voltage (think of electrons tightly packed due to the cold). On the other hand, the higher the temperature the lower the voltage.
    The cells in a module are wired in series, and thus the total module voltage is the sum of the individual cell voltages. For instance, most modules have 72 cells. Each have a voltage of 0.5 volts, so the module voltage is 36 volts.
- *Current* of a solar cell is based on solar irradiance
  - If you monitor the current of a module throughout the day you will see it fluctuates with changing irradiance.

Cold temperature sizing for string inverters

PV designers determine the number of modules in a string based on the string inverter's maximum input voltage in cold weather. If that voltage is exceeded, the inverter is shut off. Designers must also consider the string inverter's minimum input voltage in warm weather. If that voltage is *not* met, the inverter will not turn on.

It is necessary to determine the maximum number of modules that can be wired in series to a string inverter in order to avoid exceeding the inverter's maximum input voltage. When modules are wired in series, per the rules of a series circuit, the individual module voltages are summed to get the total voltage of the string.

In Figure 2.11, we have three strings of ten modules each. In each string, the modules are wired in series – from the negative terminal of one module to the positive terminal of the next. The strings are wired in parallel – the positive terminal of the first module in each string is connected in the combiner box, and the negative terminal of the last module in each string is connected in the combiner box. Wiring strings in parallel increases the current. The total current of the array is 14.85 amps (4.95 amps × 3).

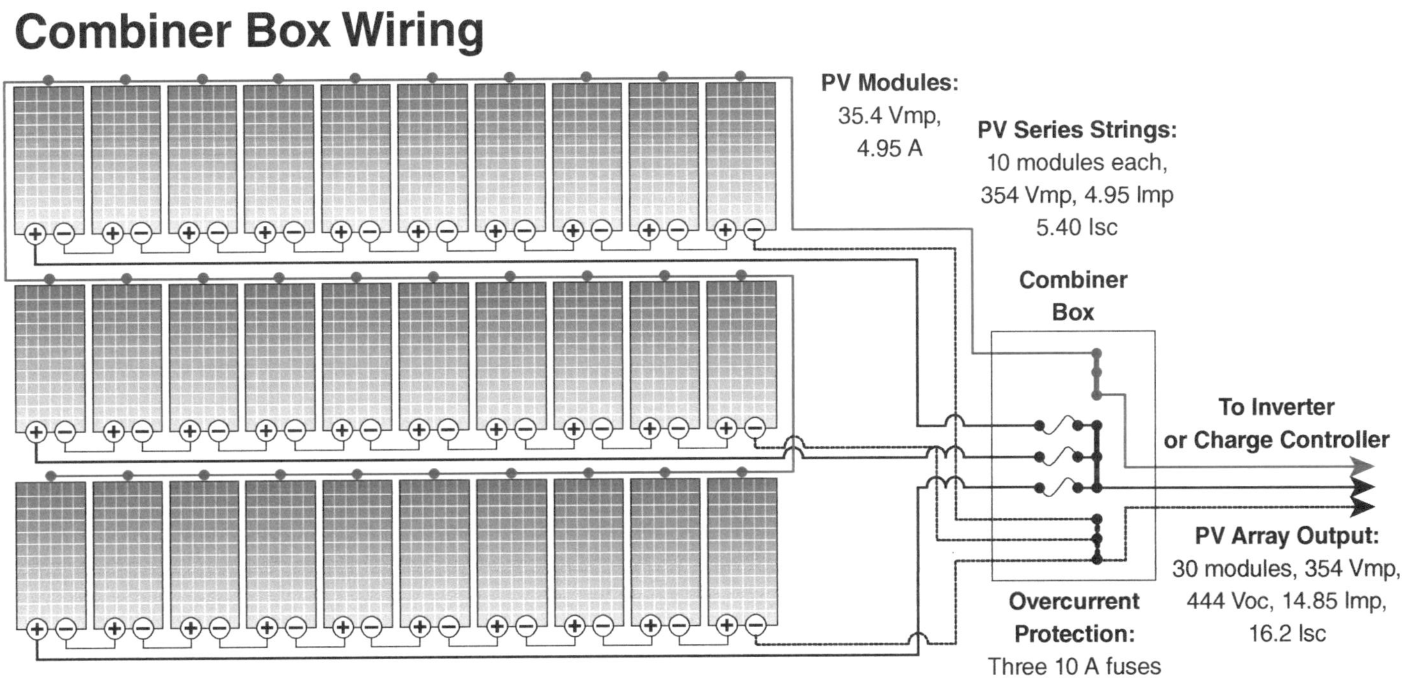

Figure 2.11 Diagram showing solar modules wired in series in a string and strings in parallel.

Source: *Homepower Magazine*. www.homepower.com/articles/solar-electricity/equipment-products/pv-combiner-box-buyers-guide.[5]

When doing wire sizing and for listing on the DC disconnect, the 2017 National Electric Code (690.8(A) Calculation of Maximum Circuit Current) requires that we use the short circuit current and multiply that by 125% to account for wild fluctuations in current due to irradiance. Thus, the maximum current (Imax) listed on the DC disconnect and for wire sizing purposes would be 5.40 Isc × 3 strings in parallel × 1.25 = 20.25 amps.[6]

Let's take an example for maximum modules in a string in cold weather: By consulting climate data from ASHRAE[7] (recommended by the NEC) or National Ocean and Atmospheric Administration (NOAA), let's say we find the coldest expected temperature in our climate over 50+ years is −15° Celsius. Module voltage is based on **standard test conditions (STC)**, which is a temperature of 25° Celsius. This gives us a difference of −40° C between the coldest temperature and STC (−15–25). The module specifies a temperature coefficient, how much the open circuit voltage will increase or decrease with the temperature. Given a typical temperature coefficient of 0.31%/C (note Kelvin and Celsius can be used interchangeably for changes), this means that for every degree higher than the STC, the open circuit voltage (Voc) will decrease by 0.31%, and for every degree lower than the STC the Voc will increase by 0.31%.

Multiplying the difference between STC and coldest temperature by the temperature coefficient (−40° C × −0.31%/C) gives us 12.4%, which is by how much the Voc will increase as a result of the cold temperature.

Given the module Voc is 39V, the increase in Voc is 12.4%, which is 1.124 (12.4/100 + 1) times 39V, and you get 43.84 Voc. This is the highest voltage that could be expected as a result of the cold temperature.

If the string inverter has a maximum voltage of 500v, then the maximum number of modules that could be placed on this circuit in series is simply 500v/43.84 Voc = 11.41 modules. Round down to 11 modules.[8] Table 2.1 lists these steps for quick reference. Module Voc and temperature coefficient can be found on the module datasheet, and maximum input voltage on the inverter datasheet.

Table 2.1 Cold temperature string sizing

Cold expected temperature = −15° C
Difference between coldest temperature and 25° C standard test conditions (STC) = (−15–25) = −40
Temperature coefficient Voc (open circuit voltage) = −40° C × −0.31%/C (from module specifications) = 12.4% increase Voc
Voc of module = 39V
Increase in Voc cold = 12.4/100 + 1 = 1.124 × 39V = 43.84 Voc
Max input voltage of inverter = 500V
Max number of modules in series = 500V/43.84 Voc cold = 11.41 modules
ROUND DOWN = 11 modules

Solar cell physics

Humans have sought to harness the sun's energy in one form or another since the dawn of time. Several scientific discoveries in the 1800s and early 1900s enabled key steps forward. As the name photovoltaic suggests, combining the word **photon**, energy from the sun, and **voltaic**, electricity, photovoltaic production can be defined as the production of electricity by light. This phenomenon, known as the **photoelectric effect**, was discovered by several scientists, including Albert Einstein.

These scientists discovered that light can essentially act as a voltage source, ejecting electrons from **semiconductors** (with a little chemical manipulation).[9] Light is in fact both a wave and a particle: a wave made from a stream of quantum particles or packets of energy called photons. Just as in the circuits described earlier in this chapter, a voltage source is required. In this case it is photons, which strike the material and are absorbed and put to use. The energy of the photons excites electrons in the outer, or **valence, shell** of the atoms that make up the material. As shown in Figures 2.12 and 2.13, it is these valence electrons that are dislodged and harnessed in a circuit.

These valence electrons absorb the solar energy and are thrust into the **conduction band**, where they are detached from their atoms to freely move around a circuit.

The voltage of a solar cell is based on the energy of the dislodged electron, known as the **photoelectron**, which is the energy of the photon minus the energy needed to dislodge the valence electron. Each photon dislodges only one electron, although advanced cells using nanomaterials allow for more

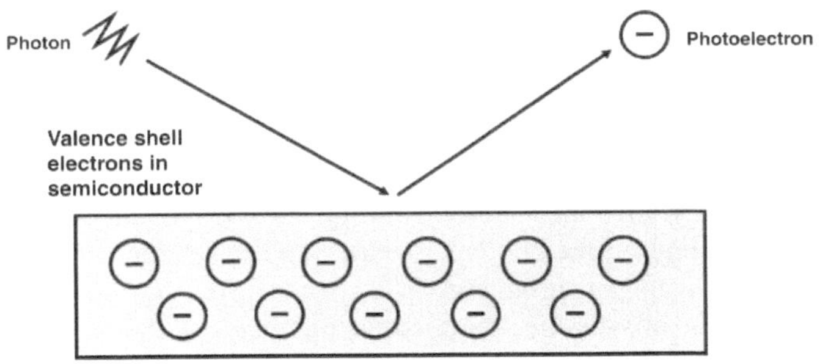

Figure 2.12 Light waves hitting the surface of a conductor and dislodging electrons to become free photoelectrons.

Source: *Harness It: Renewable Energy Technologies and Project Development Models Transforming the Grid* by Michael Ginsberg.

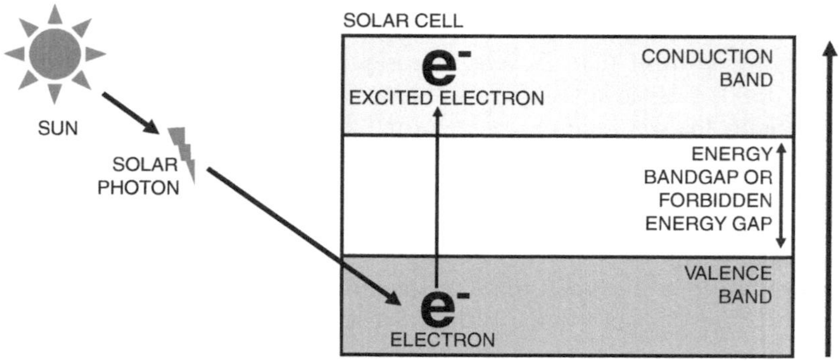

Figure 2.13 Light energy in the form of a solar photon hits the valence band within a semiconductor and excites an electron, which jumps across the "forbidden energy gap" to the conduction band to be used as current.

Source: *Harness It: Renewable Energy Technologies and Project Development Models Transforming the Grid* by Michael Ginsberg.

than one photoelectron per photon. The current of a solar cell is therefore proportional to the number of photons, or intensity of the light, reaching the cell, defined as the number of photons per second. The greater the intensity, the more photoelectrons, and so, the more current.

A cell's voltage depends primarily on the energy of the absorbed photons and the types of photons that it can absorb. A material's **energy bandgap** is the spectrum of light that it absorbs. A material that absorbs short wavelength light absorbs high energy photons and provides a high voltage. The tradeoff is that *there is less light absorbed*, which means it produces low current. That is why in Figure 2.14, voltage is shown with an arrow pointing to the left (high voltage, low current).

On the other hand, a material with a low bandgap will absorb light with short as well as long wavelengths. Thus, it creates a large current density (why current is shown with an arrow pointing to the right), but due to its low bandgap, the energy of the photons in excess of the bandgap are wasted as heat, and a low voltage is produced.

There are three possibilities for photons interacting with a material:

1 *Photons with energy less than that of the bandgap* are not absorbed.
2 *Photons with energy equal to the bandgap* are efficiently absorbed.
3 *Photons with energy greater than the bandgap* are strongly absorbed, but the excess energy is wasted as thermal energy.

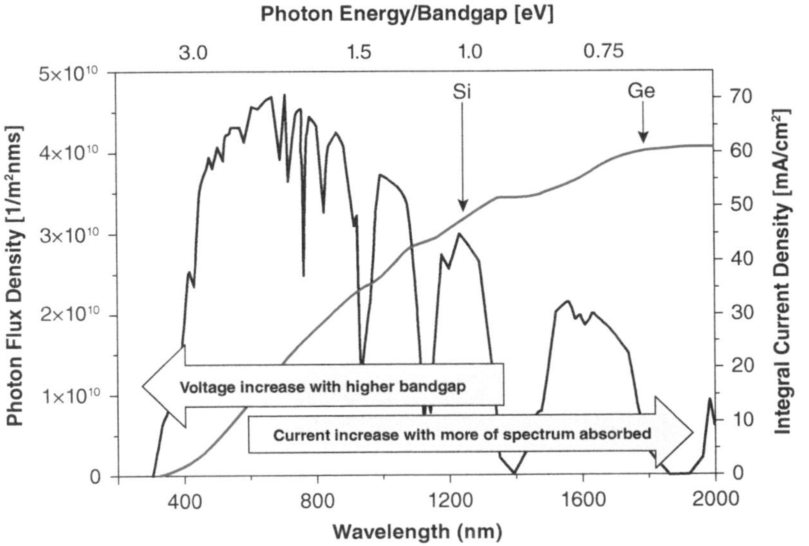

Figure 2.14 How a semiconductor's energy bandgap determines its absorption properties, and thus the voltage and current it can provide to a circuit. Photon flux density is defined as photons per square meter per wavelength interval (in nanometers) per second.

Source: Franz-Josef Haug. "Efficiency limits of photovoltaic energy conversion." www.superstrate.net/pv/limit/.

The relationship between energy bandgap (in electron volts) and wavelength is shown through the German physicist Max Planck's equation, electron volts (eV) = 1.24/lambda (lambda is the wavelength in micrometers).

Short wavelength light has high energy, resulting in a large eV. Long wavelength light has low energy, resulting in a small eV. For instance, if the lambda is 1 micrometer that results in an eV of 1.24, and if the lambda is 10 micrometers that results in an eV of 0.124.

This tradeoff between voltage and current means that a solar cell has a maximum of 31% solar conversion efficiency, known as the "Shockley-Quiesser" limit. Figure 2.15 shows various materials, their energy bandgaps and their efficiencies.

To recap, some materials absorb a little high energy light (because photons below a material's energy bandgap are not absorbed) and some absorb a lot of low energy radiation (but the extra energy in excess of the material's bandgap is wasted as thermal energy).

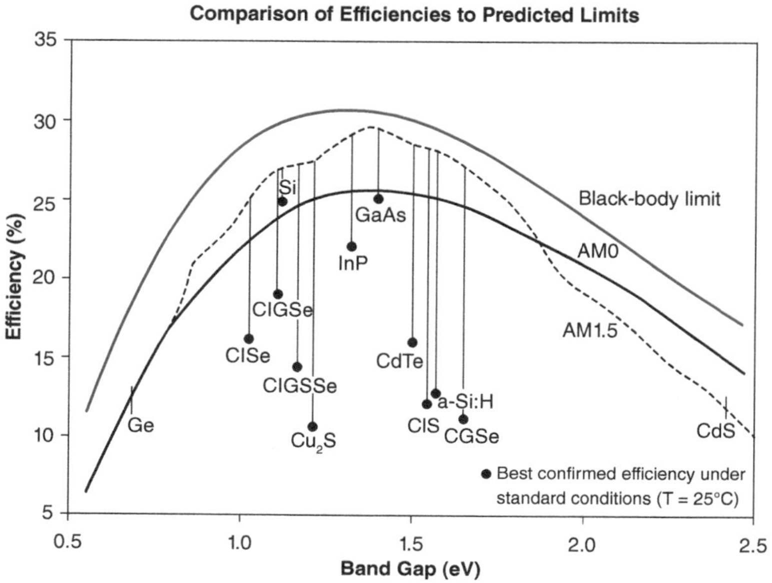

Figure 2.15 Energy bandgaps and efficiencies of different light absorbing compounds.

Source: Birkmire, R. and Kazmerski, L. *Harnessing the Sun with Thin-Film Photovoltaic*. Institute of Energy Conversion and NREL.[10]

Note: Efficiencies have increased since the year of the figure's publication (1999).

Solar cell chemistry

Silicon (Si) is the material most commonly used in today's solar cells. Si is a semiconductor, and it can conduct electrons with a little "prodding." The Si atom has four valence electrons that form **covalent**, or shared, **bonds**, making them tightly held by their atoms, as shown in Figure 2.16.

Since the valence electrons are tightly held by the atoms, we need to intentionally create imperfections in the Si crystal in a process called *doping* to allow for free electron movement. As shown in Figure 2.17, by introducing chemicals like phosphorus (P), which has five valence electrons, we create the *n-type semiconductor*. The n-type Si now has extra electrons.

We also introduce boron (B), which has three valence electrons, into Si to create the *p-type semiconductor*. The lack of electrons in this type mean the crystal has *holes* where electrons could be.

When the n-type and p-type semiconductors are put together, the extra electrons in the n-type jump into the holes in the p-type, forming the *p-n*

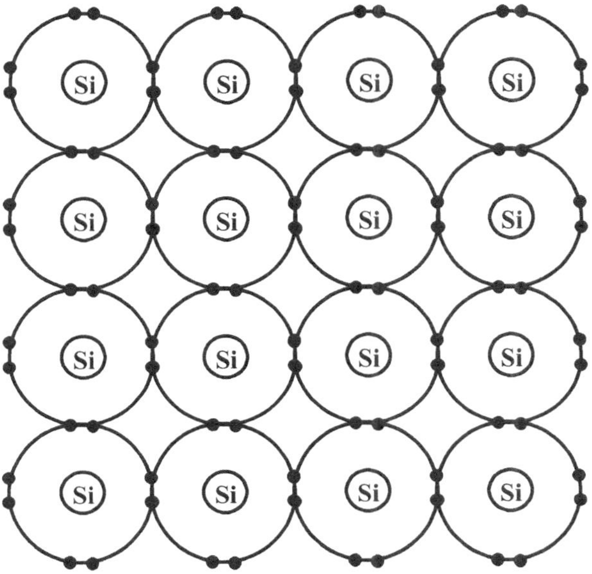

Figure 2.16 Molecular structure of silicon.

Source: *Harness It: Renewable Energy Technologies and Project Development Models Transforming the Grid* by Michael Ginsberg.

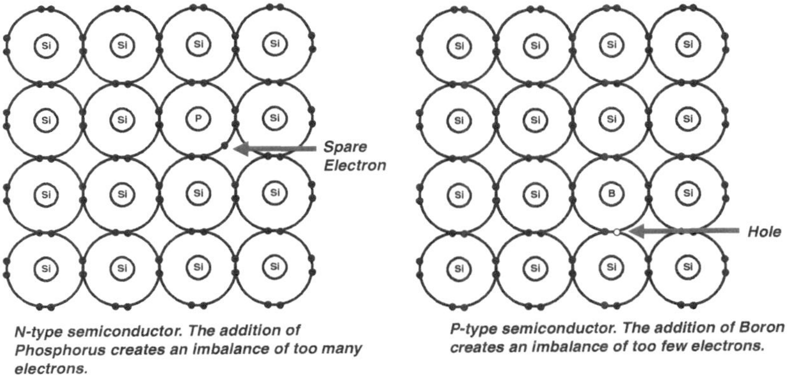

Figure 2.17 Molecular structure of P- and N-type semiconductors.

Source: *Harness It: Renewable Energy Technologies and Project Development Models Transforming the Grid* by Michael Ginsberg.

junction. In effect, the extra electrons in the p-type give it a negative charge and the holes in the n-type give it a positive charge. This creates an all-important electric field. As shown in Figure 2.18, when photons reach the p-n junction, the dislodged electrons are attracted to the positively charged n-type semiconductor, and move across the electric field, generating current which can be used in a circuit.

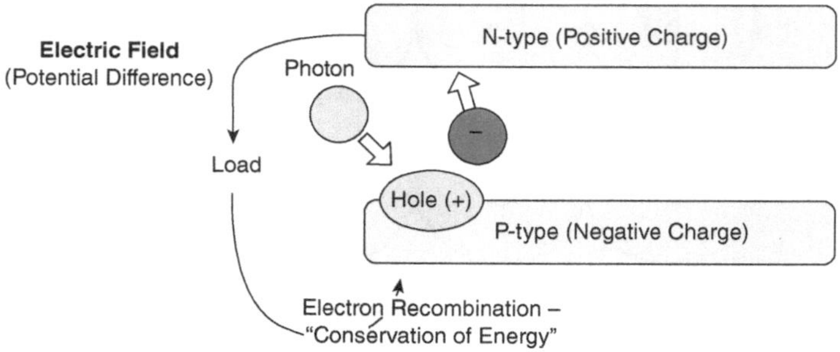

Figure 2.18 How current is generated at a P-N junction in the presence of a light source.

Source: *Harness It: Renewable Energy Technologies and Project Development Models Transforming the Grid* by Michael Ginsberg.

Solar cell components

Solar cells are made up of many layers, as shown by the simplified diagram in Figure 2.19. Back and front metallic contacts are needed to complete the electrical circuit. The n-type semiconductor is sprinkled on top of the p-type. P-type with boron is 1,000 times thicker than N-type with phosphorus. Next comes a contact grid, an antireflective coating to prevent optical loss, and a protective glass cover.

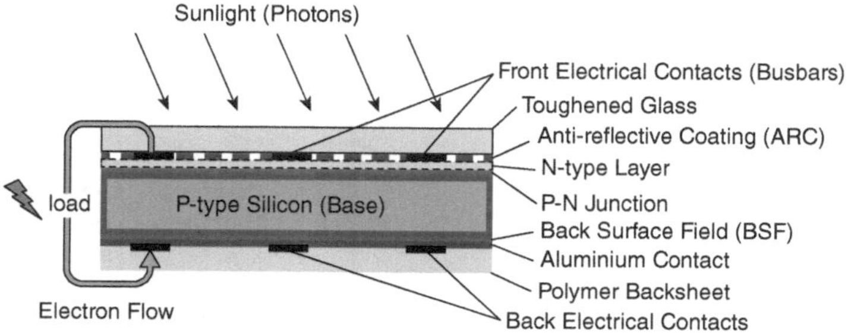

Figure 2.19 The construction of a solar cell with a p-n junction.

Source: Clean Energy Reviews (Jason Svarc).[11]

The solar module and Balance of System

As shown in Figure 2.20, solar cells are wired in series to form a module. Modules are wired in series to form a panel. Panels are wired in series and parallel to form an array. When most people describe a solar panel they are actually talking about a module! Recall from the earlier discussion, and as shown in Figure 2.21, when items are wired in series (from the positive terminal of one module to the negative terminal of the next) the voltages are summed. Panels are often wired in parallel as well to increase the current. This is all done to optimize the power output, which is equal to voltage × current.

Turning the product of **PV cells** and panels into useable electricity requires a complex network of additional components known as the **Balance of System** or **BoS**.[12]

As shown in Figure 2.22, the Balance of System needs to safely perform three functions:

1 *Condition the energy:* convert the DC power generated into AC power for consumption by common AC equipment.
2 *Transmit the energy:* conduct the electricity to the electric load/s that will consume it.
3 *Store the energy:* either charge a battery for future use, or transmit the energy to the grid for storage.

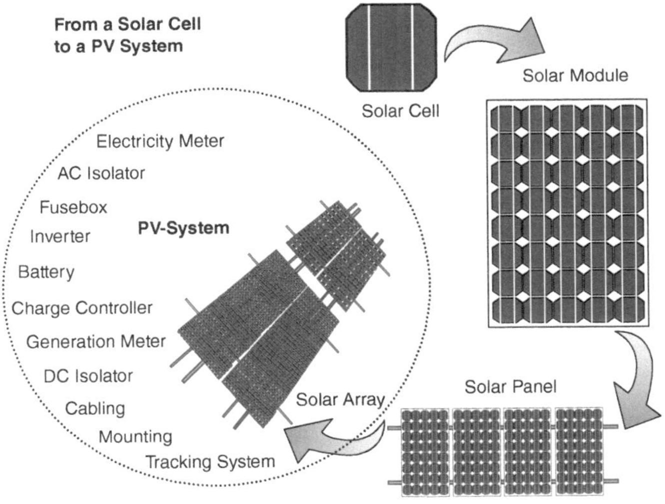

Figure 2.20 Progression from cell to module to panel to array.

Source: Wikimedia Commons. https://en.wikipedia.org/wiki/File:From_a_solar_cell_to_a_PV_system.svg.

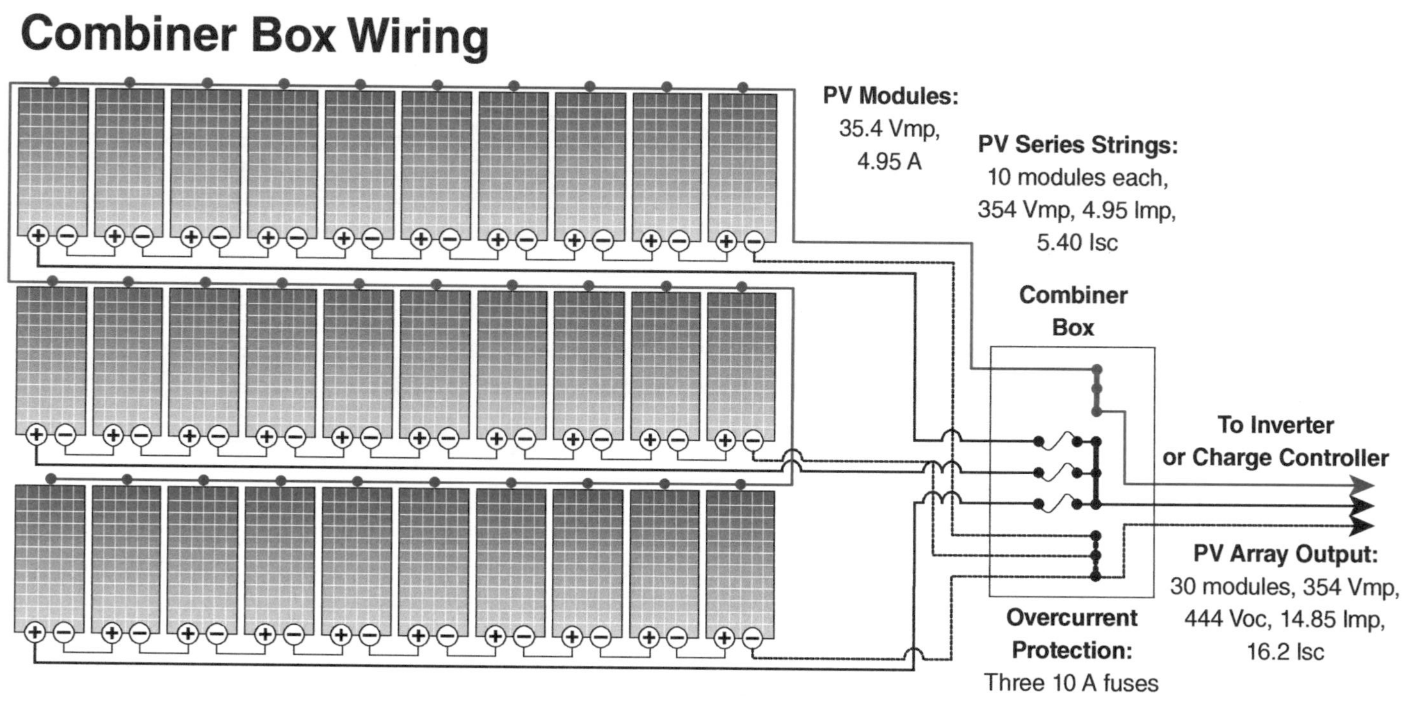

Figure 2.21 Series and parallel wiring. The modules are wired in series to form three strings of ten modules each. The three strings are wired in parallel (all the positive terminals are connected in the combiner box and all the negative terminals are connected in the combiner box).

Source: *Homepower Magazine*. www.homepower.com/articles/solar-electricity/equipment-products/pv-combiner-box-buyers-guide.

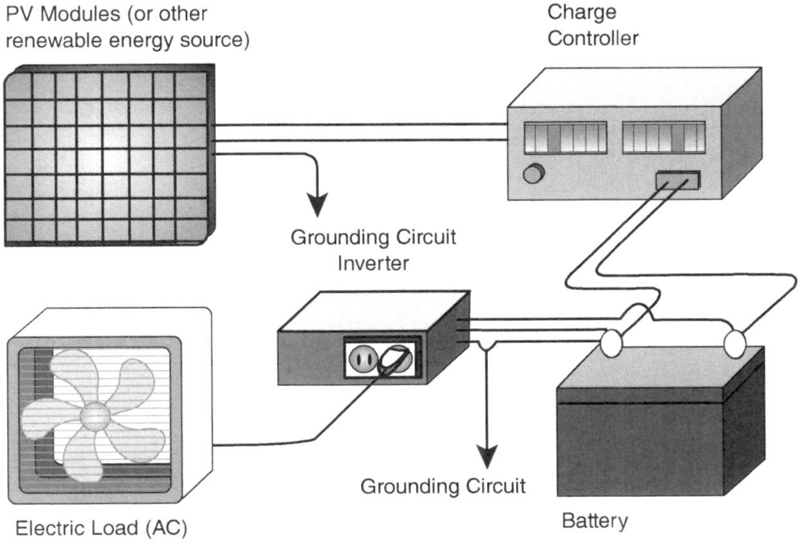

Figure 2.22 Simple Balance of System for a stand-alone PV system requiring AC power for the electrical load.

Source: U.S. Department of Energy.[13]

The specific equipment required by a Balance of System will vary by the type and need of each PV installation. The key components are outlined in Table 2.2.

Table 2.2 The key components of the PV Balance of System

Function	*Equipment*	*Use*	*Type of PV system*
Conditioning the energy	Inverters (convert DC to AC)	The DC electricity generated by the PV system is converted to AC for use by conventional electric loads. Also, inverters condition electricity to match the quality requirements of the load or the electric grid: voltage, phase, frequency, sine wave profile, etc. In addition, they capture electricity from the solar array at its maximum voltage and current, called the **maximum powerpoint (MPP)**.	All
	Charge controller	Regulate the current (amperage) from the PV system to the battery to ensure the battery does not become over-charged or over-drained.	Stand-alone and battery-backup systems
Transmitting the energy	Wiring	Conducts the generated current from the PV system throughout the rest of the Balance of System to the load. Also includes grounding for protection.	All
Storing the energy	Battery	Stores power generated by the PV system.	Stand-alone and battery backup systems
Monitoring the energy	Meters and instrumentation	Monitors battery charge and energy consumption (stand-alone) and/or power sent to and drawn from the grid (grid-tie and grid-tie/battery backup).	All
Interconnecting for large-scale commercial and utility-scale systems	**Transformer**	Steps up voltage from PV array.	Commercial and utility-scale grid-tie solar plants
Safety equipment	Overcurrent protection devices (OCPDs) (AC and DC circuit breakers); equipment grounding conductor and grounding electrode conductor, surge protection, arc fault circuit interrupter	Protects PV systems and people from short circuit currents and fires	All

Components of a solar array

The following section provides an in-depth look at how equipment in the solar system functions, providing an overview of how things should work before we analyze equipment failure in Chapter 3.

Module

In a module, PV cells are wired in series. When one cell is shaded it becomes a load for the other cells. This creates a short circuit, and current from the other cells in the string builds up, creating a **hot spot** on the bad cell. Another cause of hot spots is **cell mismatch**, in which cells of varying current production are connected in series due to errors in the manufacturing and production process. As discussed, in series circuits, the current is the same throughout the circuit and limited by the resistance of each load. Thus, the cell with the greatest resistance limits the current of the entire circuit.

Bypass diodes are now used in strings to allow for current to flow around faulty cells. A diode is a device that allows current to flow in only one direction. Bypass diodes are connected in parallel either with an individual solar cell, ideally, or series of cells, to be economical. In normal operation, the bypass diodes have no effect on the circuit because they have opposite polarity, or are *reverse biased.* As shown in Figure 2.23, similar to a check valve, reverse bias means electrons are prohibited from flowing, while flow is permitted when *forward biased.*

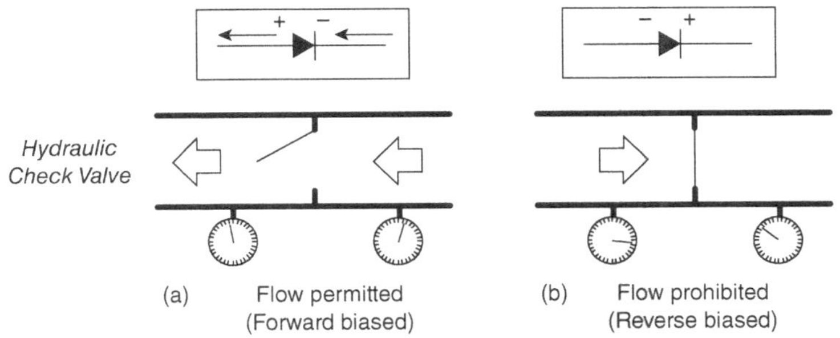

Figure 2.23 Hydraulic check valve analogy showing how electrical current flows in one direction through a diode, compared with water flowing through a pipe with a valve: (a) Electron current flow permitted; (b) Current flow prohibited.

Source: *All About Circuits.*[14]

Thus, in normal operation, solar cells are forward biased, permitting electron flow. Due to shading or cell mismatch, a cell may become reverse biased, preventing electron flow. In this instance, the bypass diode of the shaded cell conducts because the resistance is lower than that of the cell, allowing some current to flow around the bad cell, hence its name "bypass."

It is clear that in the event a bypass diode fails, the continued accumulation of current on the bad cell leads to overheating of the cell, causing the module to crack and potentially start a fire. Even with the use of bypass diodes, a faulty or poorly-producing cell (due to shading) can severely degrade the output of an entire string.

In sum, bypass diodes are wired in parallel with PV cells. When a cell is inoperable, or shaded, the electrons travel through the diode instead of the cell, as electrons will always take the path of least resistance and it is easier for the current to travel through the diode. If the bypass diode is not functional the current will remain "halted" at the faulty cell.

Shunts are alternate paths for current flow with lower resistance than the desired path. In solar cells, shunts occur at the cell edges and are due to impurities near the edges that provide a short circuit path around the p-n junction, leading to decreased performance. Shunts are one of the major factors that contribute to solar cell malfunction, and can occur at any location in the circuit where localized current is increased. Most functional solar cell errors come from a shunt and/or the series resistor.

We can use the equivalent circuitry of a solar cell to understand the effects of shunts and hot spots. As shown in Figure 2.24, when light hits the solar cell, a current is generated (I_L).

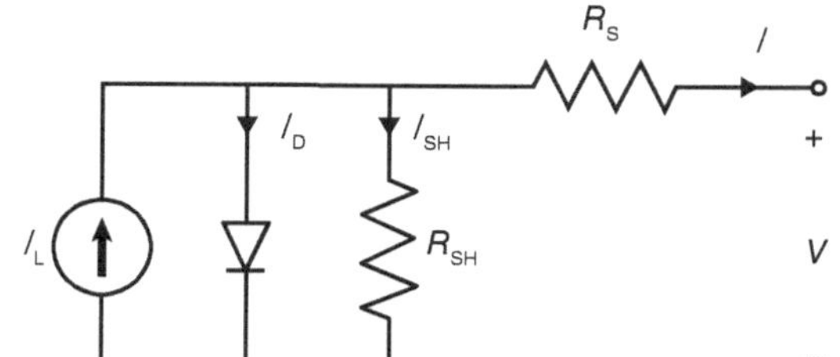

Figure 2.24 Equivalent circuit of a solar cell.

Source: Wikimedia Commons. https://commons.wikimedia.org/wiki/File:Solar_cell_equivalent_circuit.svg.

I_L is the photogenerated current
I_D is the bypass diode current
I is the output current
R_s is the series resistance – we want this to be low to allow maximum current flow
Rsh is the shunt resistance – we want this to be high (a high shunt resistance increases current density on the desired path, while a low shunt resistance decreases the useful current by providing an alternate path)
Ish is the shunt current
V is the voltage across the output terminals of the cell

The total current produced by the solar cell equals the current from the source minus diode and shunt resistor currents.

$$\mathbf{I = I_L - I_D - Ish}$$

Two conditions must be met in order to have a fully operational solar cell:

1 *High shunt resistance* (R_{sh}) – makes solar cells more efficient by increasing the current density on the desired path. Low shunt resistance from cell defects and imperfections is undesirable because it results in a loss of useful power from the cell, decreasing its efficiency and increasing its temperature since that energy is dissipated as heat.
2 *Low circuit series resistance* (R_s) – makes solar cells more efficient by increasing current density on the desired path.

Inverter

An inverter manipulates the DC electrical output of a solar array and transforms it into AC output. In DC, electrons flow in a straight path. This flow must be modified to switch (alternate) back and forth. Thus, an inverter must have a switching mechanism that flips the incoming DC repeatedly. In a basic electromechanical inverter, DC is fed into the primary windings of a toroidal (donut-shaped) transformer, which is connected to a spinning plate. As the plate rotates, the secondary winding of the transformer receives the energy as AC. Also, since there are more windings on the secondary side of the transformer, the voltage is "stepped up" from the low DC voltage of the array to the 120 volts or 220 volts of everyday AC systems.

As shown in Figure 2.25, this AC, however, is very "choppy" due to the sudden changes in direction and takes on the form of a square block wave that is not ideal for electronic equipment.

Mechanical switching was later replaced by **transistors**, which act as a switch with no moving parts (among other functions, such as amplifying weak signals). In a bipolar junction transistor (BJT), shown in Figure 2.26, two diodes are placed together. If the voltage is high enough it will permit electron flow, but if it is not high enough, then one of the diodes will become reverse biased, causing the transistor to shut off.

Transistors are used in all electronics today. Due to advances in nanotechnology, or the science of miniaturizing components to a molecular level, there are about two billion transistors in the average smart phone processor!

Capacitors, which are devices that store energy, aid in the switching of a transistor by ensuring the appropriate amount of energy is output to the transistor. For instance, in a relaxation circuit, two capacitors are charged in an

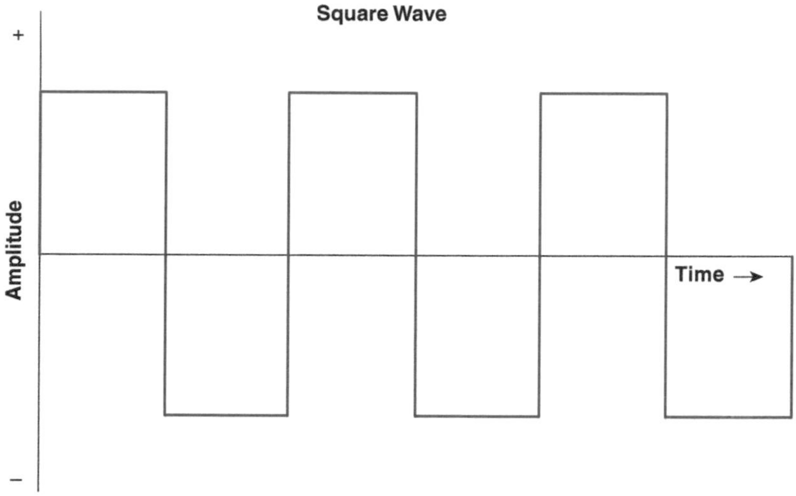

Figure 2.25 Square wave AC output from electromechanical inverter.

Source: Shawn Hymel, Sparkfun.[15]

Figure 2.26 A typical bipolar transistor.

Source: Cristian Storto/Shutterstock.com.

alternating fashion. When one capacitor discharges to the transistor, the other capacitor charges. This creates a modified sine wave, as shown in Figure 2.27.

In a modified sine wave inverter, the AC output is a rounded-off square wave, which poses issues to electronics. However, they are cheaper than pure sine wave inverters.

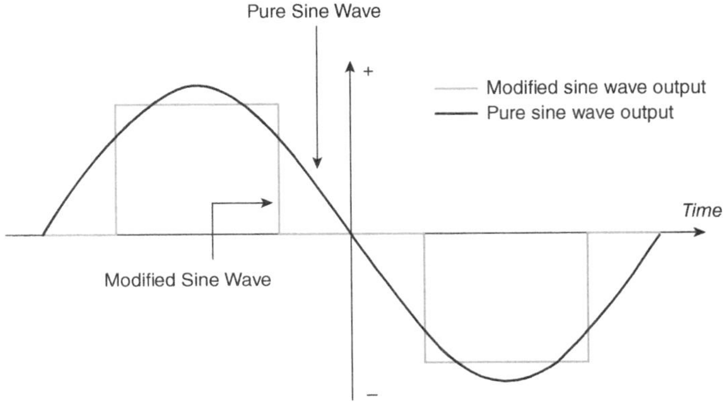

Figure 2.27 A modified sine wave compared to a pure sine wave.

Source: Photonic Universe.[16]

True sine wave inverters are better for electronics requiring certain voltages and run motors more efficiently, as motors require the current peaks not present in a modified sine wave.

Inverters smooth out the wave to produce either a modified sine wave or a pure sine wave through the use of oscillators, which feed some of the amplified output of a transistor back to the input as positive feedback that can be tuned to produce the desired frequency.

A more commonly used type of transistor today is known as a metal oxide semiconductor field effect transistor (MOSFET). MOSFET permit or restrict electron flow through the use of a gate, similar to a check valve. As shown in Figure 2.28, it has three main parts: a *source* where electrons enter, a *drain* where they exit, and a *gate*, which controls the flow from the source.

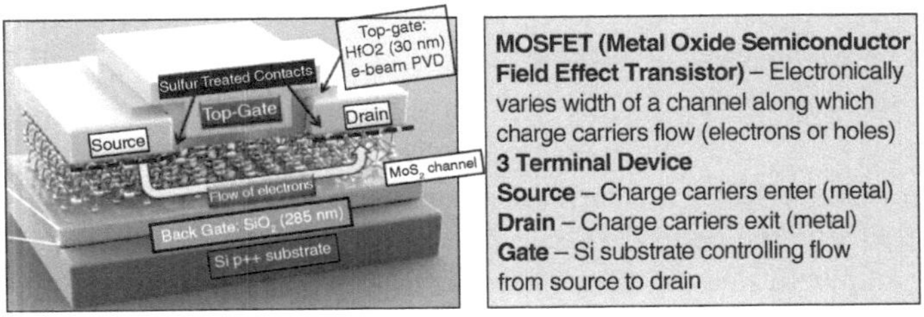

Figure 2.28 Cutaway of a metal oxide semiconductor field effect transistor (MOSFET).

Source: Bhattacharjee et al.[17]

Similarly, a thyristor acts like a transistor with the electron flow controlled by a gate, yet it has four layers of semiconductors, while a transistor has three. Thyristors essentially act like a combination of two transistors, and are able to deliver more power than transistors. Also, thyristors cannot be used as amplifiers. The insulated gate bipolar transistor (IGBT) is similar to the MOSFET but can deliver more current because of low resistance in the conducting channel. Gate failure is a potential failure mode for inverters and will be discussed more in Chapter 3.

While many inverters today use pulse width modulation (PWM), which ensures output voltage is steady regardless of the load, the best type for solar PV are known as **maximum powerpoint tracking (MPPT)** inverters.[18]

MPPT inverters optimize the output of an array. We know that power equals voltage × current and, through MPPT, the point at which the greatest power is generated is taken by the inverter, as shown in Figure 2.29. MPPT increases a module's current, and thus power output by 30% or more.

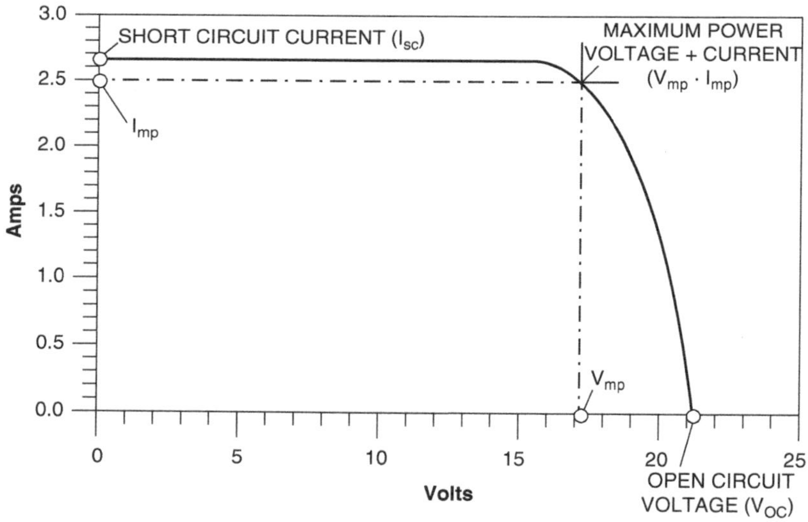

Figure 2.29 Current-voltage (IV) curve showing the maximum powerpoint (MPP) at the height of the slope.

Source: Samlex America Inc.[19]

In IV curve tracing, the current and power of each string is measured and compared with the optimal IV curve in Figure 2.29. In Figure 2.30, reduced power output can be due to a variety of issues, and through tracing the IV curve, we can begin to diagnose poor performance.

A final note: **transformerless inverters**, without the donut-shaped transformers described previously, are increasing in popularity. As opposed to

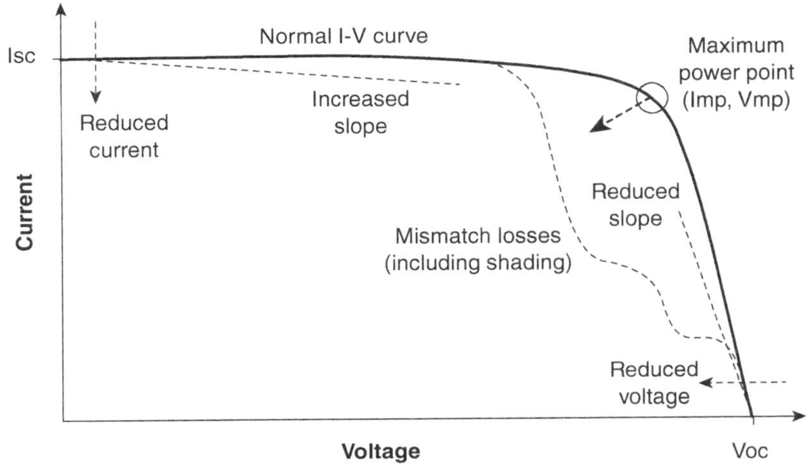

Figure 2.30 The five types of deviation from the normal IV curve shape.
Source: Solmetric.[20]

low frequency and high frequency inverters, which use transformers to step up the voltage, these inverters do not have a transformer and instead use a computerized multistep process. This reduces the weight and increases the efficiency of the inverter. These inverters allow for dual MPPT, which means strings with different solar azimuths (compass directions) and tilt angles can be connected together.

Until recently, transformerless inverters were only in Europe due to the fact that all electrical systems in the U.S. are solidly grounded. In a traditional inverter, there is *galvanic (electrical) isolation* between the DC and AC sides of the circuit through the transformer, but in a transformerless inverter there is no isolation. Case in point, before the 2017 U.S. National Electrical Code (NEC), they were known as ungrounded inverters. The transformerless inverter, however, can be grounded through additional circuitry, and are becoming more widely accepted in the U.S. In addition, transformerless inverters reduce the possibility of a ground fault since they can detect and interrupt faults as low as 300 mA and do not have a grounded conductor connected to a ground fault prevention (GFP) fuse, which could create an unintended path to ground. Since these transformers are ungrounded, there is no current flow to ground faults once the inverter has tripped. With greater knowledge of their functioning, the NEC was updated in 2017 to state a fuse is required on only one polarity (either positive or negative).

Charge controller

Similar to inverters, **charge controllers** use either PWM or MPPT to control the output of a solar array, but to a battery. Batteries are sensitive devices that require a specific input voltage and current, and get damaged when both under- and over-charged. PV power output is anything but consistent, so charge controllers are used to control and modify that output so a battery will operate under acceptable conditions.

The first charge controllers were just simple circuit breakers that tripped the circuit when the battery voltage reached a certain level. PWM monitors the battery's state of charge and sends pulses of energy from the array to the battery in varying lengths based on what is needed. MPPT controllers, on the other hand, convert excess voltage into additional current that would be wasted by a PWM controller. MPPT controllers are about 95% efficient, while PWM controllers can be as low as 65% efficient. In addition, MPPT controllers can handle higher voltages from PV arrays, which means more modules can be connected in series versus parallel for the same output and thus smaller cables can be used. They are the only types allowed for a grid-tie battery backup system where higher voltages are present. Of course, this comes at greater cost – MPPT controllers are often two to three times more expensive. Also, many new controllers come with Bluetooth capability that allows remote monitoring of the power output from the controller to the battery.

Batteries

Batteries convert chemical to electrical energy. They are comprised of two metallic terminals called electrodes, known as the anode and the cathode, a separator, and an electrolyte that facilitates the flow of charge between the electrodes. Electron flow is created from chemical interactions through a series of repeating oxidation and reduction reactions, known as redox, in which electrons are lost and gained.

Alessandro Volta developed the first battery, known as the voltaic or galvanic cell, in 1800 by alternating layers of zinc and copper separated by paper soaked in a saltwater solution. The zinc gave up electrons (oxidized) and the positive ions in the water gained electrons (reduced). As shown in Figure 2.31 of the voltaic cell, the redox cycle thus creates an electron flow between the negative anode and positive cathode.

Even though dry chemical paste (dry cells) are primarily used today, the same principle applies regardless of the material – a metal is oxidized, freeing electrons for useful work, before being gained by another substance that

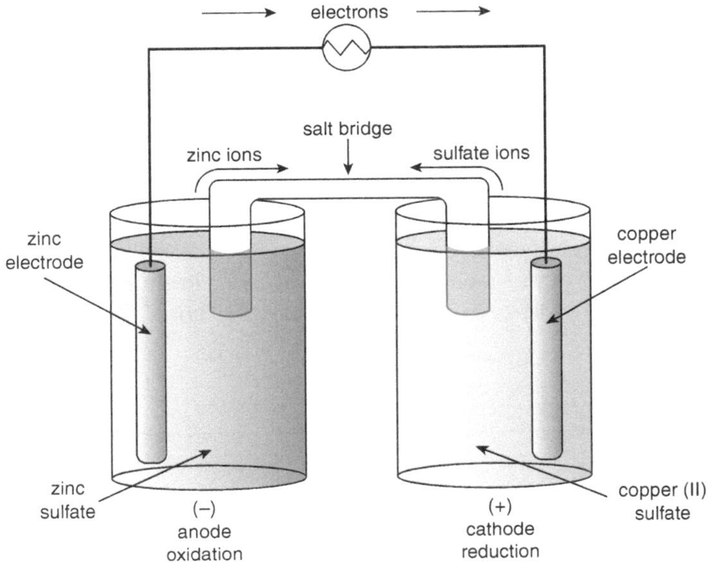

Figure 2.31 The voltaic (galvanic) cell.

Source: Wikimedia Commons. https://en.wikipedia.org/wiki/Galvanic_cell#/media/File:Galvanic_cell_labeled.svg.

is being reduced. A separator, often porous cardboard, separates the anode and cathode from direct contact.

Batteries die when all of the metal has been oxidized, but rechargeable batteries allow for the reversal of the redox reaction, which makes electrons in the metal available once more for oxidation. This process of discharging and charging is known as a cycle and the life of a battery can be measured by its cycle life, or number of cycles until the battery is dead. Ultimately, the battery will die because recharging creates imperfections in the metal's surface, which will permanently prevent oxidation.

Diverse battery technologies should be compared using several criteria, and not just cycle life. These include:

- *Depth of discharge (DoD)* – the maximum percentage of a battery's capacity that should be used, beyond which the battery's life will be shortened. A battery with a high DoD means that most of its capacity can be used, while a battery with a low DoD means it has limited capacity. A good DoD would be 95%+. Completely depleting a battery is not advised as it severely reduces its lifespan.

- *Roundtrip efficiency* – the energy output from a battery divided by the energy input to charge the battery.
- *Energy density* – the amount of energy stored in a given volume or mass.

In PV systems, which are usually rated for a 25 year+ life, it is especially important to use a battery with high DoD, roundtrip efficiency, and energy density.

Batteries for PV systems should especially have a high DoD because a PV array fluctuates during the day and there will not always be enough energy to fully charge the battery. This is the major reason why lead acid batteries are not ideal for PV systems.

Three major battery technologies – lead acid, **lithium ion**, and flow batteries – are compared across these criteria in Table 2.3.

Table 2.3 Primary battery technologies

Battery technology	*Operation*	*Uses*	*Advantages*	*Disadvantages*
Lead acid	Charge and discharge of sulfuric acid between positive and negative plates (electrodes). On the positive plate the electrolyte of concentrated sulfuric acid stores most of the chemical energy. On the negative plate sulfate ions are repelled and hydrogen ions attracted.	Vehicles, portable equipment.	Low upfront cost, mature technology, roundtrip efficiency of 80%.	Short life, low depth of discharge, low energy density, sulfation on terminals due to depletion of charge increases maintenance cost and reduces life. **Lack of charge from solar and wind can lead to sulfation and reduce life.**
Lithium ion	Lithium ions transferred between electrodes.	Consumer electronics, large-scale energy storage, solar plus storage, microgrids.	Light and high energy density, high depth of discharge (some have 100%).	High upfront cost. Roundtrip efficiency of 70–80%. Lose charge over time. Heat up and ignite at high temperatures.

Battery technology	*Operation*	*Uses*	*Advantages*	*Disadvantages*
Flow battery	Electrolytes stored in two separate tanks are pumped past a membrane.	Large-scale energy storage.	High depth of discharge (some have 100%), low maintenance. Low life.	High upfront cost. Low energy density (a lot of space required). Roundtrip efficiency of 65–75%.

Source: *Harness It: Renewable Energy Technologies and Project Development Models Transforming the Grid* by Michael Ginsberg.

- Lead acid batteries have a short life, low depth of discharge, and low energy density. Sulfation on terminals due to depletion of charge increases maintenance cost and reduces life. This means that *lack of charge from variable solar and wind can lead to sulfation and reduce life!*
- *Lithium ion batteries are light, have long lives, high energy density, and high depth of discharge (some have 100%).*

Summary

In this chapter we have learned the electricity fundamentals required to understand PV O+M and M+V. In the next chapter, we will build on this knowledge to diagnose the ways in which PV equipment can fail and how they can be fixed.

Solar PV systems are the product of increasingly-sophisticated electronics that rest on the discoveries of the past few centuries of scientists and engineers. Components continue to improve, and we are in the midst of a revolution in our understanding of nanotechnology and quantum mechanics. As our understanding deepens further, we will be able to achieve greater solar cell photoelectron efficiencies, and lengthen the life and reduce the cost of inverters, batteries, and charge controllers.

Notes

1 Ginsberg, M. *Harness It: Renewable Energy Technologies and Project Development Models Transforming the Grid*. Business Expert Press. May 2019.
2 Ohm's Law Wheel. Wikimedia Commons. https://en.wikipedia.org/wiki/File:Ohm%27s_law_formula_wheel.JPG.

3 Hukseflux Thermal Sensors. https://commons.wikimedia.org/wiki/File:SR20_pyranometer_1.jpg. This file is licensed under the Creative Commons Attribution 4.0 International license.

4 SolarGIS. Solar Resource Maps. https://solargis.com/maps-and-gis-data/overview/.

5 *Homepower Magazine*. www.homepower.com/articles/solar-electricity/equipment-products/pv-combiner-box-buyers-guide.

6 Note conditions of use, such as temperature, impact the wire size determination.

7 ASHRAE Climate Data Center. www.ashrae.org/technical-resources/bookstore/ashrae-climate-data-center.

8 See NEC 690.7(A)(1), voltage temperature calculation method, and Sean White's *Solar Photovoltaic Basics: A Study Guide for the NABCEP Entry Level Exam* for more examples.

9 Conductors cannot be used because the electrons are too free to create an electric field.

10 Birkmire, R. and Kazmerski, L. *Harnessing the Sun with Thin-Film Photovoltaic.* Institute of Energy Conversion and National Renewable Energy Laboratory. www.nrel.gov/docs/fy99osti/29582.pdf.

11 Svarc, J. Solar Panel Construction. www.cleanenergyreviews.info/blog/solar-panel-components-construction. August 20, 2018.

12 "Balance-of-System Equipment Required for Renewable Energy Systems." Department of Energy. Accessed September 15, 2018. www.energy.gov/energysaver/balance-system-equipment-required-renewable-energy-systems.

13 Ibid.

14 "Introduction to Diodes and Rectifiers." www.allaboutcircuits.com/textbook/semiconductors/chpt-3/introduction-to-diodes-and-rectifiers/.

15 Hymel, S. "Alternating Current (AC) vs. Direct Current (DC)." https://learn.sparkfun.com/tutorials/alternating-current-ac-vs-direct-current-dc/alternating-current-ac.

16 DC to AC Inverter. Photonic Universe. www.photonicuniverse.com/en/how-to-choose/ac-inverter/.

17 A sub-thermionic MoS2 FET with tunable transport. *Applied Physics Letters* 111 (2017):163501. https://doi.org/10.1063/1.4996953.

18 "MPPT Algorithm." MathWorks. https://www.mathworks.com/solutions/power-electronics-control/mppt-algorithm.html.

19 "Solar Panel Characteristics." Samlex America Inc. www.samlexsolar.com/learning-center/solar-panels-characteristics.aspx.

20 "I-V Curve Tracing Exercises for the PV Training Lab." Solmetric. http://resources.solmetric.com/get/I-V-Curve-Tracing-Exercises-for-the-Outdoor-PV-Training-Lab.pdf.

3 Tools and methodology for maintenance and troubleshooting

Safety first

In order to ensure that work is done safely and securely, the proper equipment and procedures must be in place before maintenance can begin. Precautions can be divided into two categories: **personal protective equipment (PPE)** and **procedures/work practices**. The saying "an ounce of prevention is worth a pound of cure" applies here – sound work practices are best, while PPE is a last measure against harm.

The use of PPE is based on a risk assessment of potential hazards encountered while working on a PV system.

Risk = Hazard × Likelihood

In a risk assessment, the severity of each potential hazard is determined on a scale of acceptable to unacceptable risk by examining the likelihood of the hazard and the degree of its consequence. The level of risk can be determined by navigating to the appropriate box in a risk matrix like the one shown in Figure 3.1. For instance, if the likelihood is 3 and the consequence is 5, the

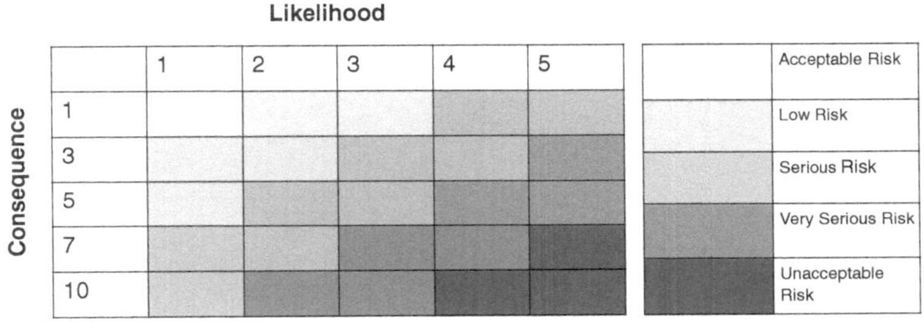

Figure 3.1 Risk matrix.

Source: Author.

Table 3.1 Risk assessment for a commercial PV installation

Hazard	*Likelihood*	*Consequence*	*Evaluation*	*Procedural control measure*	*PPE control measure*
Being hit by falling, flying or fixed objects	Average (3)	Extremely high – serious injury/ death (10)	30 = **Very serious risk**	Safety watch	Protective helmet (hard hat); safety glasses; steel-toed boots; fabric gloves
Shock from energized DC conductors near PV array	Average (3)	Very high – electrocution (7)	21 = **Very serious risk**	Lock Out Tag Out; non-conductive ladder (with fiberglass rails); use insulated tools	Protective helmet; safety glasses
Electrocution by energized AC conductors in main electrical panel (medium to high voltage)	High (5)	Extremely high – serious injury/ death due to electrocution/arc flash (10)	50 = **Unacceptable risk**	Lock Out Tag Out; non-conductive ladder (with fiberglass rails); use insulated tools	Face shield; non-conducting pole/rope; rubber gloves, sleeves, shoes; wear cotton; no loose clothing; protective helmet
Electrocution by live overhead power lines	High (5)	Extremely high – serious injury/ death due to electrocution/arc flash (10)	50 = **Unacceptable risk**	Stay at least 10 feet from a power line; non-conductive ladder (with fiberglass rails)	Face shield; non-conducting pole/rope; rubber gloves; protective helmet

Falling from a height that is 6 feet or more above a lower level	High (5)	Extremely high – serious injury/ death (10)	50 = **Unacceptable risk**	**Fall prevention**: guardrails, safety netting	**Fall protection**: body harness (personal fall arrest); rescue plan
Dehydration/heat stress/stroke/skin cancer in hot dry climates	Moderately high (4)	Very high – serious injury/death (10)	40 = **Unacceptable risk**	Safety watch; bottled water; periodic breaks	Sunglasses; sunscreen; lightweight long-sleeved SPF light-colored shirts and pants; knee pads; hats with broad brims and cape for shoulders (jackets for mornings); sneakers (unless carrying objects > 25 lbs)
Hypothermia in cold climates	Moderately high (4)	Very high – serious injury/death (10)	40 = **Unacceptable Risk**	Safety watch; periodic breaks	Layered clothing – compression smartwool top and socks with outer rainproof jacket with ventilation; insulated hat; gloves; boots

Source: Template adapted from A. Dembski.

risk would be "serious risk." There is a degree of subjectivity in this process, but determinations are made based on experience and expertise in the field.

Consider a risk assessment for a commercial PV installation in Table 3.1. The control measures reduce the respective hazard to an acceptable, or low, risk.

While all risks should be mitigated, the benefit of a risk assessment is that hazards can be categorized by risk severity. The assessment carried out in Table 3.1 results in the risk spectrum shown in Figure 3.2.

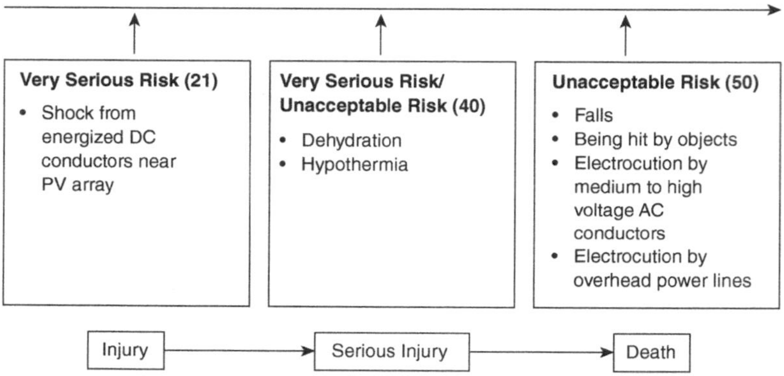

Figure 3.2 Risk spectrum for commercial PV installation.

Source: Author.

In the construction industry overall, the most common causes of death in 2017 are called the "fatal four."[1] These include, in order of likelihood:

1 Falls (381/971 deaths, or 39.2%).
2 Being struck by objects (80, or 8.2%).
3 Electrocutions (71, or 7.3%).
4 Being caught between objects (50, or 5.1%).[2]

Humans conduct electricity. The average resistance of skin is 600 ohms, although it can decrease in water. Should you unwittingly become part of a circuit, current may flow through you!

As shown in Figure 3.3, per Ohm's Law, 370 milli-Amps would flow through you if subjected to 220 volts. As little as 75 milli-Amps across the heart can kill you.

Electrocution

- Human body is about 600Ω
- R_H = 600Ω V = 220v
- 10 or 12 gauge wire ⟶ ~1.5Ω
- Short circuit (sc) ⟶ 0Ω

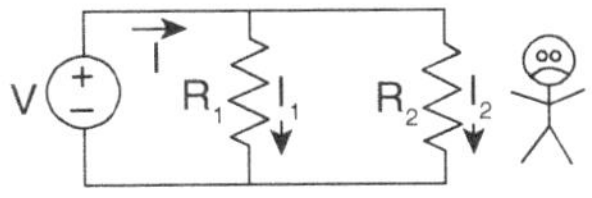

- $I_1 = V/R_1$ = 220v / 1.5Ω = 147a
- $I_H = V/R_H$ = 220v / 600Ω = .37a = 370mA
- $I_{sc} = V/R_{sc}$ = 220v / .01Ω = 22,000 amps*220v = 4.84 MW ⟶ **Arc Flash**

> ➤ Short circuit from unintended contact of components is **high amperage, low to zero resistance.**
> ➤ The branch with the **least resistance** has the **largest current.**
> ➤ **Short circuit current** flows to **ground** or, if no ground, **to humans or sensitive conductive equipment.**

75mA *across heart* can stop the heart
100mA will cause your muscles to contract

Figure 3.3 Risk of electrocution.

Source: Author.

Procedural control measures and work best practices

Implementing the following control measures and work best practices mitigates risk to keep you and the PV array you are working on safe.

Safety precautions always feel like a lot of work until the day you wished you had completed them. Don't put yourself at risk, make these work practices part of your routine. When safety is second nature, it doesn't seem a chore- you are less likely to forget a step, and are more likely to stay safe every day.

Safety watch – put simply, never work alone. Ensure that you always work with at least one other technician who can help in the event of an emergency.

Lock Out Tag Out – one of the most important work practices for PV technicians, **Lock Out Tag Out (LOTO)** is a way for workers to protect themselves and others from potentially hazardous energy and communicate with others the work that is being performed. As the name implies, the worker *locks out* a system and then *tags out* that system. A lock ensures that whatever is being worked on is off (de-energized – i.e. a circuit breaker is open). A tag is used to communicate with technicians and others who is performing the work, and what work is being performed. Figure 3.4 shows a LOTO pouch and a locked out electrical cabinet.

Figure 3.4 Top – Electrical Lock Out Tag Out pouch kit (Accuform STOPOUT®). Bottom – Lock Out Tag Out of an electrical cabinet (showing a hasp, lock, and tags in use).

Source: SocoXbreed/Shutterstock.com.

Depending on the work performed, technicians may be either "authorized" personnel or "affected" by others' work. It is the responsibility of the authorized employee and the manager to ensure the LOTO is performed and successfully communicated to everyone in the vicinity.

Although steps for LOTO are tailored to the facility/equipment, the following general steps can be applied to all situations involving electrical equipment.

Box 3.1 Lock Out Tag Out steps

Shutdown steps

1. Prepare and announce the shutdown, and state the magnitude of energy.
2. Perform machine/equipment shutdown (at disconnect or breaker feeding machine).
   - For a PV system, open the AC disconnect first and lock and tag. Next, open the DC disconnect and lock and tag.
3. Perform machine/equipment isolation.
4. Perform Lock Out Tag Out device application.
5. Release stored energy – bring equipment to a zero mechanical state (important for inverters).
6. Verify isolation.

Startup steps

1. Remove tools, reconnect lines, put machine guards back in place.
2. Announce that equipment is being activated, and ensure employees are in position.
3. Remove LOTO devices (LOTO devices can be removed ***only by the technician that applied them***).
   - First, remove the lock and tag of the DC disconnect and close. Then, remove the lock and tag of the AC disconnect and close.
4. Restore energy and test equipment.

Box 3.2 PV system safety precautions

It is important to take account of the following extra hazards and safety precautions when working on PV systems.

- During the day, and even on overcast days, modules are energized and present a potential shock hazard.
- At night, PV modules are generally not energized and present minimal hazard from electrical shock. However, scene lighting, low ambient light, or other artificial light sources can generate enough current to pose a shock hazard at night. The same safety precautions taken during the day should be taken at night.
- Never walk or climb on PV modules. Although the modules will withstand some weight, they still present a significant safety hazard due to the potential for breaking glass, tripping, and slipping. Exposure to the cells inside a module present a potential shock hazard.
- Never place roof or ridge ladders on or against the PV modules/arrays due to the potential for broken glass and resulting shock hazard.
- Never break a PV module with an ax or other forcible entry tool due to the potential for broken glass and resulting shock hazard.
- Never cut metal conduits or wires strung between PV modules or wires coming from a series of PV modules to a combiner box. This could result in serious or fatal injury from electrical shock.
- Never attempt to remove fuses from solar PV fuse boxes. (Not all PV systems have fuse boxes.) Doing so has the potential to start a fire and presents a significant electrical shock hazard!
- Securing the main electric power coming into the building only shuts down power inside the building and does not stop PV modules from producing DC power when sunlight or a light source is present. An electric shock hazard still exists from the array to the DC side of the inverter.

Source: Adapted from the San Francisco Fire Department solar PV system safety and fire ground procedures.[3]

Overcurrent protection devices (OCPDs) – **overcurrent protection devices** can be grouped by the purpose they serve: devices protecting equipment, and devices protecting people.

Grounding and bonding protects equipment and people. Grounding is the practice of directing stray electricity toward the ground through large conductors staked into the ground. The deeper you go into the earth, the less resistance, so it is ideal for the grounding rod to be drilled as deep as possible (the National Electrical Code requires a minimum depth of 8 feet).[4] Figure 3.5 shows how the grounding electrode conductor (GEC) connects to the grounding rod (electrode). The GEC is connected to the equipment grounding conductor (EGC), which bonds the modules and all conductive material in the system (the EGC is not shown).

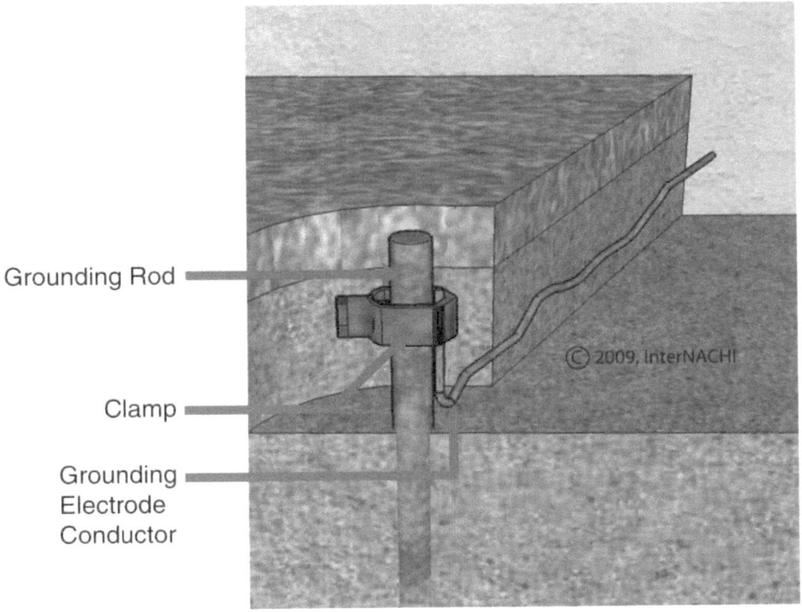

Figure 3.5 Grounding electrode conductor connected to grounding rod or electrode.

Source: Nick Gromicko and Kenton Shepard. www.nachi.org/grounding-electrodes.htm.

Bonding is the practice of connecting all metallic equipment to ensure all conductive material is connected and any stray voltage will go to ground, rather than through a person!

Fuses and circuit breakers are designed to protect equipment. Ground fault circuit interrupters (GFCIs), on the other hand, open much more quickly and with detection of less stray current are designed to protect people. Stray current indicates a short circuit, or unintended path for the current. As such, GFCIs work by sensing the current going into an outlet through the hot wire and the current exiting the outlet through the neutral wire. If there is a difference as little as 5 mA, depending on the GFCI, the device will trip to minimize the amount of time the current will course through your body.

Fuses can only be used once, while circuit breakers, which use an electromagnet to open a circuit, can be reused. Figure 3.6 illustrates the differences between fuses, circuit breakers, and GFCIs.

EQUIPMENT	PEOPLE
DANGER: Short Circuit→Overheating→Fire **PROTECTION VIA** ➢ Fuse ➢ Circuit Breaker	**DANGER:** Electrocution **PROTECTION VIA** ➢ Ground Fault Circuit Interrupter (GFCI)
GROUNDING AND BONDING – directs stray voltage *away* from equipment/people to ground	

FUSE

- In a short, current ↑↑.
- P = VI – wires heat up rapidly.
- Heat up and break before wires.
- Take long time to activate and **can only be used once.**

CIRCUIT BREAKER

- As current increases, electromagnet gets stronger and pulls down lever, disconnecting **moving contact.**
- **Reusable.**

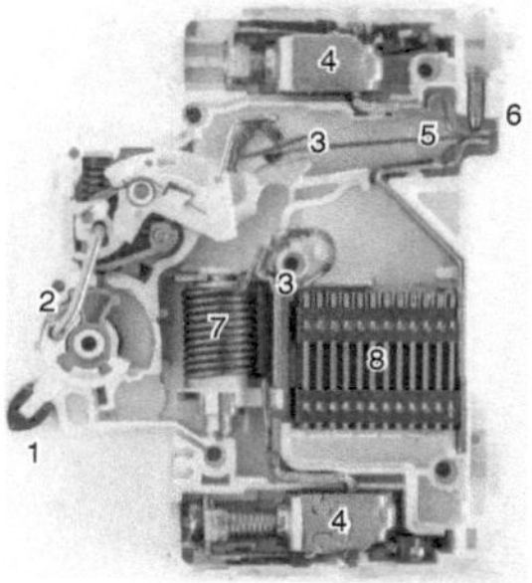

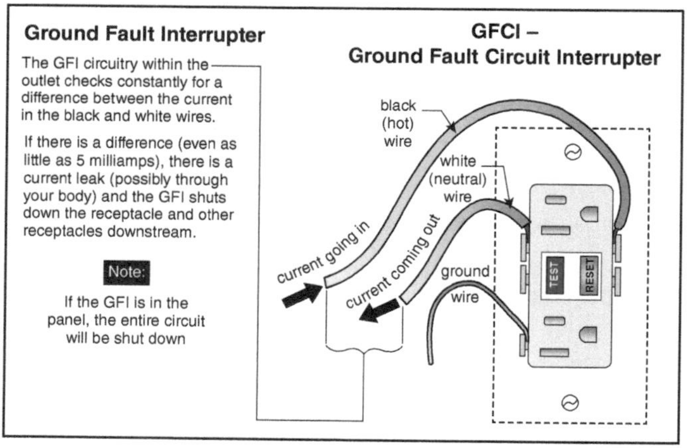

Figure 3.6 Fuse vs circuit breaker vs ground fault circuit interrupter. Fuses and circuit breakers trip a circuit to protect equipment, while ground fault circuit interrupters (GFCIs) disconnect a circuit much quicker (less than 80 milliseconds) to protect people.

Source: GFCI Image courtesy of Carson, Dunlop & Associates Ltd. © All rights reserved. www.carsondunlop.com. Fuses (Oleg Bezrukov & Ian Grainger/Shutterstock.com). Circuit Breaker (tarczas/Shutterstock.com).

Legend for the circuit breaker:

1 Actuator lever – manually trips the circuit breaker. When the circuit is tripped it is in the Off position, and when the circuit is energized it is in the On position. Most breakers are designed to trip even if the lever is held or locked in the On position.
2 Actuator mechanism – forces the contacts together or apart.
3 Contacts – permit the flow of current when touching and breaks the current when not touching.
4 Terminals – ensure a tight connection with the wires.
5 Bimetallic strip – separates contacts during small long-term overcurrent. It is made up of two metals (one with a high melting point, the other low). The one with the low melting point is closest to the electrical input and bends away from the input contact terminal to cut off electrical supply. About 10 minutes after the supply is cut off, the metal cools and comes back to normal position.
6 Calibration screw – allows the manufacturer to precisely adjust the trip current of the device after assembly.
7 Solenoid – separates contacts quickly during high overcurrent. The solenoid is an electromagnet, which has a pulling force that increases with current. The solenoid's pull releases the latch that normally keeps the contacts closed.
8 Arc divider/extinguisher – when an arc, an "electrical spark," occurs this component lengthens the arc into a chute and into the dividers to dissipate it.

A PV system has both DC and AC disconnects. In addition, the 2017 NEC Article 690.12 increased requirements for *rapid shutdown.* Rapid shutdown requires that the system voltage be reduced quickly to reduce shock hazards for technicians and firefighters. The rules differ within and outside the array boundary, which is defined as 1 foot from the array. According to the NEC 2017 690.2, the array is "a mechanically integrated assembly of module(s) or panel(s) with a support structure and foundation, tracker, and other components, as required, to form a dc or ac power-producing unit." Conductors up to 3 feet from building penetration are also considered within the array (see Figure 3.7).

Within the array boundary, modules and exposed conductive parts must be able to be reduced to 80 volts within 30 seconds. Outside of the array boundary, conductors must be limited to 30 volts within 30 seconds of rapid shutdown. Rapid shutdown devices, either at the service disconnect or a special switch, and module level power electronics (MLPE) that reduce voltage, comply with this requirement. Systems with no exposed conductive parts or where the array is installed more than 8 feet from exposed grounded conductive parts are not required to comply.[5]

Safety barriers – use these to prevent passage to a hazardous area, accompanied by signs identifying the type of hazard.

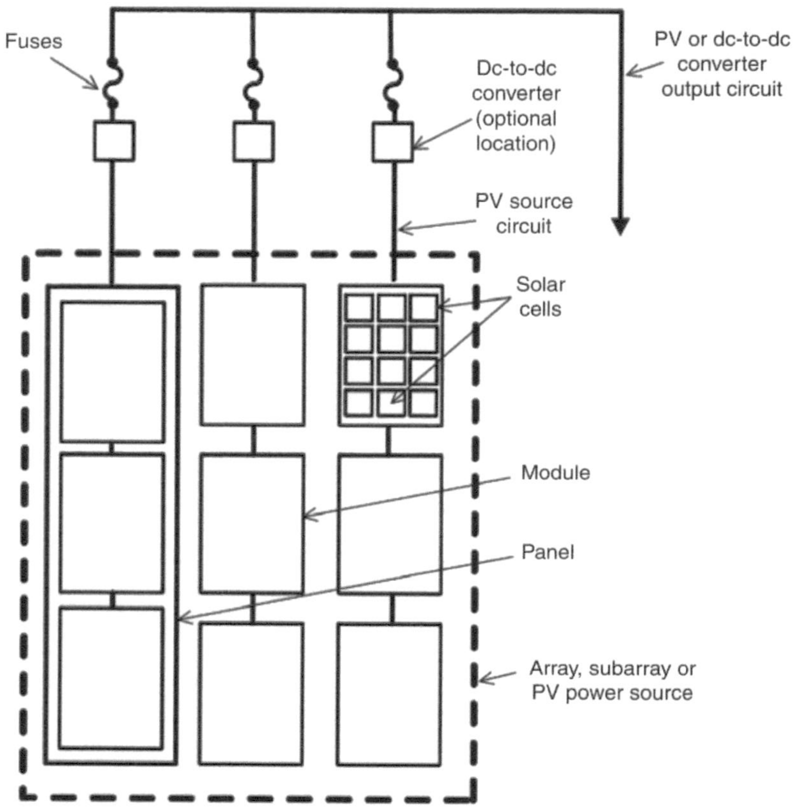

Figure 3.7 2017 NEC PV Figure 690.1(a) PV power source. The dotted square is considered the array boundary. The boundary is typically the edge of the PV modules but can be 1 foot from the edge of rails, trackers, or concrete foundations that extend beyond the modules.

Source: Reproduced with permission from NFPA 70®-2017, National Electrical Code®, Copyright © 2016, National Fire Protection Association, Quincy, MA 02169. This reprinted material is not the complete and official position of the NFPA on the referenced subject, which is represented only by the standard in its entirety, which can be obtained through the NFPA web site at www.nfpa.org.

Guard rails – install vertical rails with the top rail 42 inches +/− 3 inches (107 cm) above the walking/working level and capable of supporting a force of at least 200 pounds (91 kg). See Figure 3.8.

Safety nets – safety nets are required when the workspace will be more than 25 feet (7.6 meters) above the ground or water surface. Safety nets

Figure 3.8 Guard rails.

Source: FallProof Systems LLC.[6]

must extend 8 feet (2.4 meters) beyond the edge of the work surface where employees are exposed, and shall be installed as closely underneath the work surface as practical but in no case more than 25 feet (7.6 meters) below such work surface. See Figure 3.9.

Figure 3.9 Safety net.

Source: FallProof Systems LLC.

Ladder safety – appropriate ladder protocols are required to reduce the risk of trips and falls, per Figure 3.10.

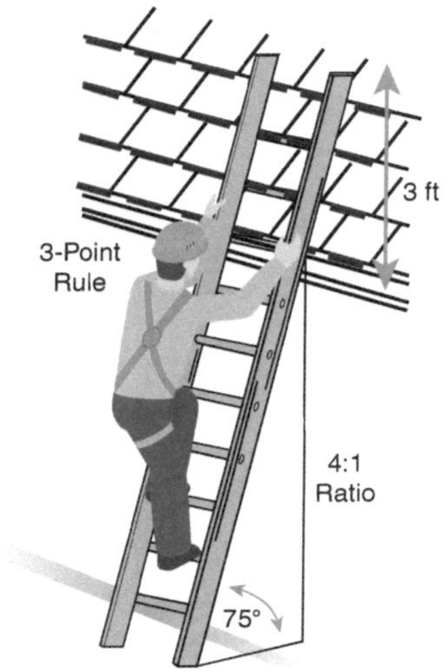

Figure 3.10 Correct ladder use.

Source: OSHA.

- 1:4 Rule – set the ladder 1 foot out for every 4 feet up to the point of support (30 cm out for every 1.2 meters up). This is easy to estimate as each rung of a ladder is about 1 foot (30 cm).
- Ensure the top of the ladder is at least 3 feet above contact with the roof.
- Observe the ladder for cracks before beginning work.
- Choose the correct height ladder for the job.
- Maintain three points of contact on the ladder at all times (e.g. two hands and one foot, two feet and one hand).
- Ensure unbroken anti-slip safety feet at the base of the ladder.
- Use a board on soft/uneven earth to stabilize the ladder.
- Do not overreach while on the ladder.
- Do not stand on the top rung of the ladder.
- Secure ladder feet to prevent movement and slippage.
- Keep ladder clean and clear.
- **And always use the RIGHT TOOL for the RIGHT JOB!**

In the picture on the left in Figure 3.11, the ladder is too short for the job of placing lights on the tree, and the worker is standing on the top rung. Also, the worker only has two points of contact with the ladder, not three (two feet, no hands on the ladder). This is unsafe because the likelihood of him falling has increased significantly! The picture on the right shows a ladder of the correct height for the job.

Figure 3.11 Use the ladder of the correct height for the job.

Source: Author and Tim Stufft.

Other tips:

- Use insulated tools.
- Cover boxes and panels.
- Use rubber mats.

Personal protective equipment (PPE) control measures

In the U.S., the Occupational Safety and Health Administration (OSHA) approves the application of PPE by work task. PPE should fit the individual and rated to protect against hazards that may be encountered while work is performed.

Protective helmets – hard hats are classified by type and class:

Type I – protects the top of the head
Type II – protects the top and sides of the head
Class G (General) – protects the head from up to 2,200 volts
Class E (Electrical) – protects the head from up to 20,000 volts
Class C (Conductive) – no protection against conductors

Technicians working on PV arrays both on the ground and the roof should wear Type II and Class G. For technicians working on arrays with step up transformers where medium voltage may be encountered, Class E helmets should be used. As shown in Figure 3.12, the type and classification can be found on the inside of the hard hat:

Safety glasses – eyewear should be comfortable and fitted to form a protective seal to prevent dust and debris from accessing the eye. Prescription

Figure 3.12 Label of certification found inside a hard hat (Type 1 Class E).

Source: MSA – The Safety Company.[7]

lenses should be incorporated into the safety glasses or prescription glasses should be worn under safety goggles.

OSHA states eye and face protection must comply with the American National Standards Institute (ANSI) Z87.1–1989 standard, although the most recent update is 2015. For PV technicians, protective eyewear should be impact rated and also offer protection from optical radiation (visible light, glare), and ultraviolet (UV) radiation. UV 400 means 100% UV protection, as shown in Figure 3.13.

Footwear – steel-toed boots are typically worn when carrying items 25 pounds (11.3 kg) or more, as shown in Figure 3.14. Sneakers with grip and ventilation should be worn when working on a roof in hot weather. Rubber boots are required when working on medium- to high-voltage systems.

Personal fall arrest system (PFAS) – OSHA requires workers to wear a body harness when exposed to the risk of falling 6 feet or more from

Figure 3.13 Eye and face protection showing ANSI Z87.1+ and UV 400 rating.

Source: Global Vision Eyewear Corp (Wind-Shield A/F with Yellow Tint Lens).[8]

Figure 3.14 Appropriate footwear with good grip and steel-toed boots. Use ventilated boots when working on a roof in hot weather.

Source: Author.

an unprotected side or edge. A PFAS consists of several components: an anchor, body harness, lifeline, shock-absorbing lanyard, and suspension trauma safety straps (stirrups).

As shown in Figure 3.15, the anchor must be able to hold 5,000 pounds (2,268 kg) and be secured to the roof with an anchor plate at the "anchor point" (1). The anchor is connected to the "body harness" (2) through large D rings on both the anchor point and the harness. A shock-absorbing lanyard ("Connector" (3)) connects the harness with the "lifeline" (4), and reduces the force experienced during a fall. In addition, the use of suspension trauma safety straps, shown in the image on the right, is recommended to minimize suspension trauma, which is the loss of consciousness due to pressure on the femoral artery, trapping blood in the legs.[9]

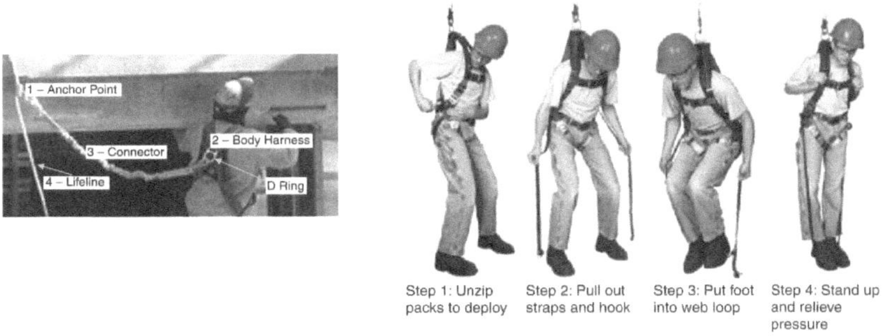

Figure 3.15 Appropriate body harness, anchor points, and suspension trauma safety straps.

Source: WeeklySafety[10] and 3M.[11]

Clothing safety tips – workers should wear cotton clothes, have no metal in their clothing, and no loose clothing.

Arc flash PPE

An **arc flash** occurs when medium to high voltage electricity travels through the air between conductors, generating an explosion at extremely high temperatures (four times hotter than the surface of the sun). Additional PPE is required to provide adequate protection against a potential arc flash.

Face shield – technicians authorized to work on terminals with medium to high voltage AC must wear face shields. These face shields are rated by the amount of discharged heat energy that it can withstand in cal/cm^2.

Long clothing/rubber gloves – technicians must wear specialized clothing and rubber gloves rated by the calories/cm^2 that it can withstand. Figure 3.16 shows the PPE for arc flash hazards by the energy it can withstand. PPE is rated by the severity of the potential arc flash and must be able to withstand a minimum cal/cm^2 per category.

Figure 3.17 shows an electrician wearing category 4 arc flash PPE with a minimum arc rating of 40 cal/cm^2.

Non-conducting pole/rope – in the event of an emergency, technicians will use a non-conducting pole or rope to pull a colleague safely away from energized equipment.

PPE Category Level Chart NFPA 2009		Minimum Arc Rating of PPE (Calories/cm^2)
CATEGORY 0	What Personal Protection Equipment (PPE) You Shall Wear: [X] Cotton Undergarments [X] Long Sleeved Shirt (Natural Fiber) [X] Long Pants (Natural Fiber) [X] Safety Glasses or Goggles [X] Hearing Protection (Inserts) [X] Leather Gloves (as needed) or Insulating Gloves with Protectors 	0
CATEGORY 1	What Personal Protection Equipment (PPE) You Shall Wear: [X] Cotton Undergarments [X] Arc Rated Long Sleeved Shirt (or FR Coveralls) [X] Arc Rated Long Pants (or FR Coveralls) [X] Hard Hat with Arc Rated Face Shield [X] Hearing Protection (Inserts) [X] Safety Glasses or Goggles [X] Leather Gloves or Insulating Gloves with Protectors [X] Leather Shoes (as needed) 	4
CATEGORY 2	What Personal Protection Equipment (PPE) You Shall Wear: [X] Cotton Undergarments [X] Short Sleeved "T" Shirt (Natural Fiber) [X] Arc Rated Long Sleeved Shirt and Long Pants or Arc Rated Coveralls Instead [X] Hard Hat with Arc Rated Face Shield with Sock Balaclava [X] Safety Glasses or Goggles [X] Hearing Protection (Inserts) [X] Leather Gloves or Insulating Gloves with Protectors [X] Leather Shoes (as needed) 	8
CATEGORY 3	What Personal Protection Equipment (PPE) You Shall Wear: [X] Cotton Underwear [X] Short Sleeved "T" Shirt (Natural Fiber) [X] Arc Rated Long Sleeved Shirt and Long Pants [X] Arc Rated Coveralls (over the above) [X] Arc Rated (25 cal) Arc Flash Suite Jacket [X] Arc Rated (25 cal) Arc Flash Suite Pants [X] Arc Rated (25 cal) Arc Flash Suite Hood [X] Hard Hat [X] Safety Glasses or Goggles [X] Hearing Protection [X] Arc Rated Leather Gloves or Insulating Gloves with Protectors [X] Leather Shoes 	25
CATEGORY 4	What Personal Protection Equipment (PPE) You Shall Wear: [X] Cotton Underwear [X] Short Sleeved "T" Shirt (Natural Fiber) [X] Arc Rated Long Sleeved Shirt and Long Pants [X] Arc Rated (40 cal) Arc Flash Suite Jacket [X] Arc Rated (40 cal) Arc Flash Suite Pants [X] Arc Rated (40 cal) Arc Flash Suite Hood [X] Hard Hat [X] Safety Glasses or Goggles [X] Hearing Protection [X] Arc Rated Leather Gloves or Insulating Gloves with Protectors [X] Leather Shoes 	40
Always wear voltage rated rubber glove liners when working above 50 volts in all categories!		

Figure 3.16 Categories of PPE required for each arc flash category.

Source: OEL Worldwide Industries.

Figure 3.17 Electrician wearing arc flash PPE while working on an energized circuit.
Source: Jay Petersen/Shutterstock.com.

Now review all of the safety measures and PPE discussed in this chapter, and take a look at Figure 3.18. Can you identify what the individual is doing well and what he is missing?

Figure 3.18 Rooftop PPE dos and don'ts in a PV system installation.

Maintenance methodologies and best practices

"Look for the horse, not the zebra." Or if you're in Tanzania, "Look for the zebra, not the horse." The point is, look for what is most likely to fail first before spending time on less-likely failures.

From preventive maintenance to reliability-centered maintenance

The idea behind an operations and maintenance (O+M) plan is to save money by extending the life of a system. **Preventive maintenance (PM)** via regularly scheduled repairs has been shown to reduce equipment failure and unplanned emergency repair time that would otherwise be incurred by relying on **reactive maintenance** (waiting until something breaks to fix it). Another benefit of PM is that it actually reduces planned repair hours.

However, from a cost perspective, it is possible to do too much PM, and it can end up being counterproductive. **Reliability-centered maintenance (RCM)** advocates for a more sophisticated, equipment-specific O+M plan that takes into account equipment criticality levels, the impact of a failure, and potential failure modes. Figure 3.19 displays this spectrum from reactive maintenance to RCM.

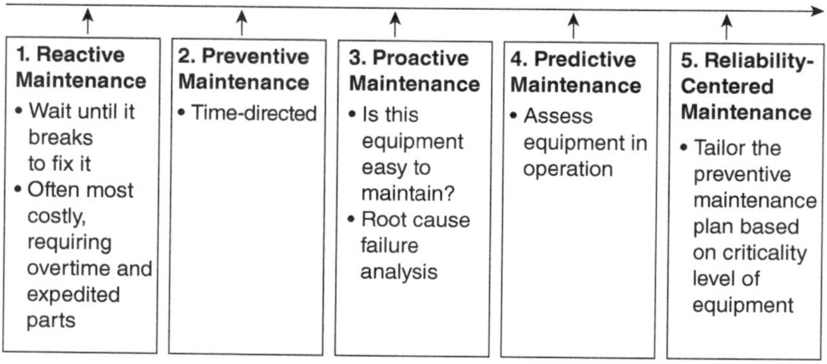

Figure 3.19 The maintenance spectrum.

Source: Author.

Failure mode and effects analysis for PV system

In a **failure mode and effects analysis (FMEA)**, a technician assesses each component of a system, and each of the potential ways in which it can fail. Then, the *risk priority number (RPN)* of the failure mode is determined by multiplying the *severity* of a failure with the *probability* that it will fail, and with the *detectability* or likelihood that a control is in place to detect the potential failure before it fails. Severity, also known as consequence, is based on the degree to which a failure, by order of importance, would endanger human health or jeopardize equipment operation. Legends for each factor are shown in Figure 3.20.

Effect	SEVERITY of Effect	Ranking
Hazardous without warning	Very high severity ranking when a potential failure mode affects safe system operation without warning	**10**
Hazardous with warning	Very high severity ranking when a potential failure mode affects safe system operation with warning	**9**
Very High	System inoperable with destructive failure without compromising safety	**8**
High	System inoperable with equipment damage	**7**
Moderate	System inoperable with minor damage	**6**
Low	System inoperable without damage	**5**
Very Low	System operable with significant degradation of performance	**4**
Minor	System operable with some degradation of performance	**3**
Very Minor	System operable with minimal interference	**2**
None	No effect	**1**

PROBABILITY of Failure	Failure Prob	Ranking
Very High: Failure is almost inevitable	>1 in 2	**10**
	1 in 3	**9**
High: Repeated failures	1 in 8	**8**
	1 in 20	**7**
Moderate: Occasional failures	1 in 80	**6**
	1 in 400	**5**
	1 in 2,000	**4**
Low: Relatively few failures	1 in 15,000	**3**
	1 in 150,000	**2**
Remote: Failure is unlikely	<1 in 1,500,000	**1**

Figure 3.20 Legends for severity, probability, and detectability.

Source: FMEA Worksheet (adapted from A. Dembski).

Detection	Likelihood of DETECTION by Design Control	Ranking
Absolute Uncertainty	Design control **cannot** detect potential cause/ mechanism and subsequent failure mode	**10**
Very Remote	**Very remote** chance the design control will detect potential cause/mechanism and subsequent failure mode	**9**
Remote	**Remote** chance the design control will detect potential cause/mechanism and subsequent failure mode	**8**
Very Low	**Very low** chance the design control will detect potential cause/mechanism and subsequent failure mode	**7**
Low	**Low** chance the design control will detect potential cause/mechanism and subsequent failure mode	**6**
Moderate	**Moderate** chance the design control will detect potential cause/mechanism and subsequent failure mode	**5**
Moderately High	**Moderately high** chance the design control will detect potential cause/mechanism and subsequent failure mode	**4**
High	**High** chance the design control will detect potential cause/mechanism and subsequent failure mode	**3**
Very High	**Very high** chance the design control will detect potential cause/mechanism and subsequent failure mode	**2**

Figure 3.20 (Continued)

The potential failure modes of each piece of equipment are then ranked by their RPN, from highest to lowest. The higher the RPN, the more critical that an action be taken to lower the risk. For each potential failure mode, risk is mitigated by taking an action to reduce one or more of the preceding factors to come up with a lower RPN, one that is more acceptable for the situation. Although a degree of subjectivity is involved in determining an RPN, in-depth knowledge of the operation of equipment is essential to making an accurate assessment.

In the following section, each component of a solar PV Balance of System is analyzed for its potential failure mode, with a focus on detectability/ diagnostic techniques and action measures. While climate and maintenance procedures vary across sites, the following FMEA assumes a temperate climate that is not overly dusty, and standard maintenance procedures. Some more sophisticated procedures, which are employed in large-scale solar farms, are also included. The goal is to reduce the criticality of the potential failure mode to the acceptable region, as shown in Table 3.2.

Table 3.2 Levels of criticality and associated risk priority number

Criticality		*Risk priority number*
Most critical		> 150
Critical		100 ≤ 150
Significant		75 ≤ 100
Acceptable		< 75

Now we will perform an FMEA for each component in a PV system, as shown in Figure 3.21.

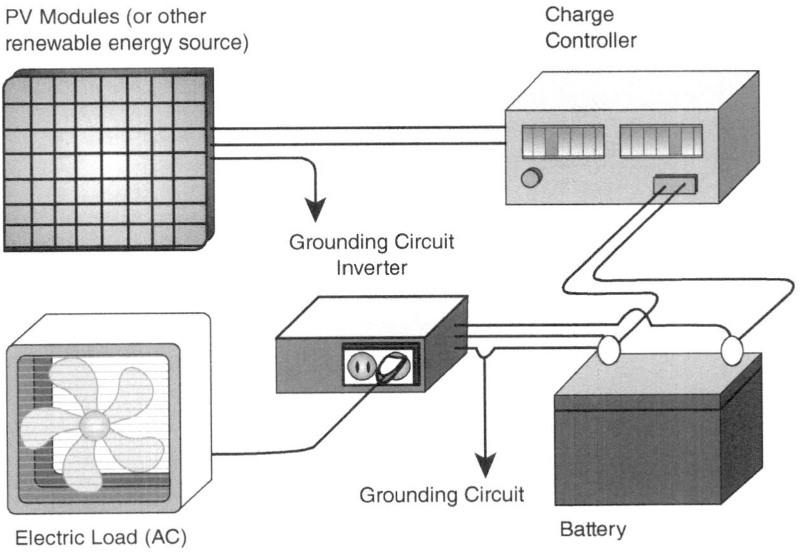

Figure 3.21 Simple Balance of System for a stand-alone PV array requiring AC power for the electric load.

Source: U.S. Department of Energy.[12]

Solar Module FMEA part 1: diagnostic techniques and risk priority number

Potential failure mode/effect of failure/cause of failure	*Severity*	*Probability*	*Detectability*	*Detectability/diagnostic technique*	*Risk priority number (RPN)*
Failure mode: Bypass diode failure and hot spots *Effect*: Reduced power output and fire *Cause*: Uneven soiling See Figure 3.22 for example of infrared aerial imaging to identify solar cell and module abnormalities	**9** – Hazardous with warning	**5** – Moderate	**6** – Low (will be detected if power monitoring plan in place and infrared imaging done)	Infrared camera, visual or aerial thermal imaging, power monitoring to include current transformers (CTs) measuring current from combiner box home runs at inverter, and/or back of module temperature sensor	**270** (9 × 5 × 6)
Failure mode: Cracked backsheet, frame, and glass *Effect*: Reduced power output *Cause 1*: Installers or maintenance technicians standing on module *Cause 2*: Water infiltration and ribbon corrosion See Figure 3.23 for example of cracked module glass	**7** – High (system inoperable with equipment damage)	**6** – Moderate	**3** – High (*assuming power monitoring or aerial imaging in addition to visual inspection*)	Visual or aerial thermal imaging, power monitoring to include current transformers (CTs) measuring current from combiner box home runs at inverter, and/or back of module temperature sensor	**126** (7 × 6 × 3)
Failure mode: Soiling *Effect*: Reduced power output *Cause 1*: Dusty or sandy climate See Figure 3.24 for example of dust on PV modules *Cause 2*: Bird droppings See Figure 3.25 for example of bird droppings on modules	**4** – Very low (system operable with significant degradation of performance)	**8** – High (*highly dependent on your climate but commonly occurs*)	**3** – High (*assuming weekly inspection*)	Visual inspection, (*aerial thermal imaging via drones is used for large plants*)	**96** (4 × 8 × 3)

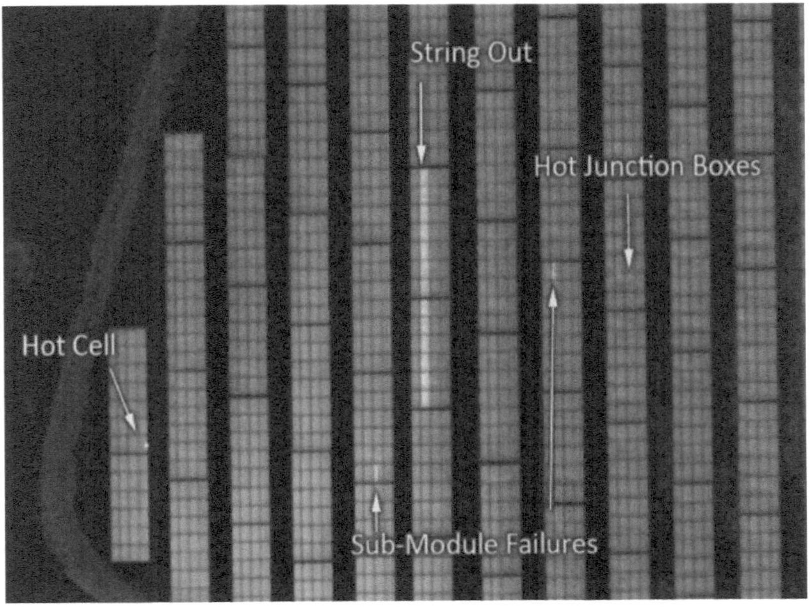

Figure 3.22 Infrared aerial imaging identifying failed junction boxes, strings, modules, and cells within modules.

Source: Rob Andrews, Heliolytics Inc.

Figure 3.23 Cracked module glass.

Source: NREL. *Best Practices in PV O+M.*[13]

Figure 3.24 Dust on solar photovoltaic module.

Source: Dachochai Saytangjai/Shutterstock.com.

Figure 3.25 Bird droppings on modules.

Source: Andy Walker. NREL. *Best Practices in PV O+M.*[14]

Detection and diagnosis of potential module failures

Module degradation rates vary by type of semiconductor material and environment. Major causes of degradation include oxygen contamination in the bulk of the Si junction (quick initial degradation) and ultraviolet exposure (long-term decline). The most commonly-installed material today is multicrystalline silicon (mc-Si), which has a degradation rate of 0.64% per year, as shown in Figure 3.26.[15] For instance, this means that for a 10 kW mc-Si system we can expect a reduction of 0.064 kW per year (10 kW × 0.0064).

PV Module Type	Degradation Rate per Year (%/year)
Amorphous silicon (a-Si)	0.87
Monocrystalline silicon (sc-Si)	0.36
Multicrystalline silicon (mc-Si)	0.64
Cadmium telluride (CdTe)	0.40
Copper indium gallium diselenide (GIGS)	0.96
Concentrator	1.00

Figure 3.26 Representative PV module degradation rates.

Source: Photovoltaic Degradation Rates – An Analytical Review, D.C. Jordan and S.R. Kurtz. NREL.[16]

Thermography is used to detect cracked backsheets, glass, and solar cell issues due to hot spots and other issues. An infrared (IR) camera, AKA thermal imager, detects and focuses the IR radiation emitted from a solar panel to create a thermogram, or temperature pattern. Hot elements typically indicate unwanted heat loss or damage. Figure 3.27 shows an electrician using a thermal imager on an electrical cabinet. Figure 3.28 shows thermograms of damaged solar cells.

Figure 3.27 Thermal imager showing the thermal characteristics of an electrical enclosure.

Source: Reproduced with permission, Fluke Corporation.

Figure 3.28 Thermograms showing detection of solar cell issues due to shunts and other causes. Hot emitting panel sections indicate damage to the solar cell.

Source: FLIR Systems.

Box 3.3 Infrared camera procedure

1 Verify the PV array is operating by checking current on each string and inverter display for instantaneous kW output.
2 Set thermal imaging camera to "Auto-Scaling" and color palette to Iron or Rainbow.
3 Point lens onto the active cells without shading it.
4 Keep no greater than 10 feet from the module and lens perpendicular to the cell.
5 Log module temperature, time, date, location in array, and picture.

Source: Adapted from *PV System Operations and Maintenance System Fundamentals*, solarabcs.org.[17]

Thermography can be used to detect multiple issues with solar cells, as shown in Table 3.3.

Table 3.3 Table showing how thermography can be used to diagnose different types of issues with a solar cell

Problem types	*Example*	*Thermal imaging appearance*
Processing induced defects	Impurities	Hot spot (or cold spot)
	Cracks inside	Crack-like heating
Damage	Cracks	Crack-like heating
	Cracks inside	Partially heated
Shading	Pollution, bird droppings, humidity	Hot spot
Bypass diode		Patchwork pattern
Bad interconnection	Not connected	Whole area hotter

Source: Chen, J. Iowa State University. "Evaluating thermal imaging for identification and characterization of solar cell defects."[18]

Solar Module FMEA part 2: corrective actions and new risk priority number

Potential failure mode	*Recommended action(s)*	*New severity*	*New probability*	*New detectability*	*New risk priority number (RPN)*
Bypass diode failures and hot spots	• Following thermographic scan, replace faulty plug-in diodes. If the diodes are soldered in or the junction boxes are sealed, replace module(s). • Select backsheets with inner layers (especially important for rooftop systems, where temperatures can be 40° C above the ambient air). • Remove shading. • Increase cleaning frequency. • Use microinverters that can detect module-level losses.	**2** – Very minor – system operable with minor interference (if all recommended actions taken)	**3** – Low	**3** – High	**18** (2 × 3 × 3)
Cracked backsheet and glass	• Repair glass with UV resistant two-part epoxy. Test performance after application. If no improvement, replace module. • For backsheet, first check if crack has penetrated the encapsulant. If not, lightly sand the backsheet and apply epoxy. Test performance after application. If no improvement, replace module. • Installation and O+M control to prevent accidental damage. • Include inspection of backsheets in regular inspections. • Ongoing power output monitoring plan.	**3** – Minor (system operable with some degradation of performance)	**3** – Low	**2** – Very high	**18** (3 × 3 × 2)
Soiling	• Use sensor to detect soiled glass. • Wash with plain demineralized (distilled) water with mild detergent. • Use robotic cleaner.	**4** – Very low (system operable with significant degradation of performance)	**3** – Low	**1** – Almost certain	**12** (4 × 3 × 1)

Corrective actions for potential module failures

Repairing cracks – after cleaning the modules, mix the UV resistant two-part epoxy. Use a scale to measure – the mixing ratio should be 10:1 resin:catalyst (hardener) – and mix into a third container. Once mixed, you have a working time of 10–15 minutes. Pour the mixed epoxy onto the glass and spread it. Do not push it over the edge of the module. Leave the module for a few hours and let it set.

Use an epoxy that does not yellow or degrade from exposure to the sun, heat, or cold. See products specifically designed for solar modules, such as "Cell Guard" from ML Solar.[19]

Array washing – wash the array using deionized water to prevent corrosion and calcium buildup on the frame. In addition, use demineralized water since frequent washing with high mineral content water may eventually leave a hard-to-remove film of mineral deposits.

Soiling can reduce array output by up to 20%. However, there is a tradeoff between cleaning and lost output. While more regular cleaning minimizes lost output due to dirt, it also increases O+M costs. Finding the optimal schedule is location-specific.

NREL recommends that facilities determine an "optimal cleaning interval" based on:

- The cost of cleaning (\$ per m^2).
- The rate of dirt accumulation, shown as power loss (% per day, month or year).
- The quality of the solar resource or the *capacity factor*, (% of actual over potential capacity).
- The value of the delivered power in \$ per kWh.
- The PV module efficiency (the lower the efficiency the more area that needs to be cleaned for the same benefit).[20]

Figure 3.29 shows a sample analysis of solar array cleaning frequency for minimal cost. In this case, cleaning three to four times per year results in minimal cost. This will vary by site. It is clear there is a tradeoff between lost output due to soiling, and the cost of cleaning and lost output due to cleaning downtime.

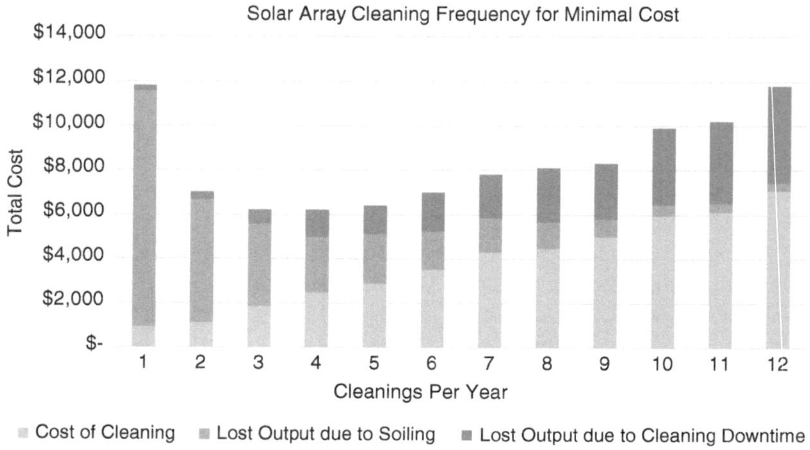

Figure 3.29 Sample site analysis of solar array cleaning frequency for minimal cost.
Source: Author.

Case study: a tale of two PV washings in West Africa

"The rain cleans them" – the penalty of dirty solar panels in Dakar, Senegal

The single most important regular maintenance activity for solar PV systems is simply washing the modules with water. As discussed, the power output and reliability of a solar module is greatly dependent on the irradiance, or quantity of photons, received on the material per second.

This means that any obstruction, be it shading, soiling, or dirt, has the potential to dramatically reduce the panel's power output. Recall that power equals voltage × current. A soiled panel significantly decreases the current, and thus power.

On a trip to a 2 MW PV array in Dakar, Senegal, the author discovered the facility team was unsure as to why their brand new (less than one year old) roof-mounted PV array was producing roughly half of its expected output. This diligent team had carefully inspected designs, investigated the inverter, and contacted the installer, thinking some software error or equipment failure had occurred.

Looking at its poor performance and eroding economic return, an already skeptical management team was quick to ridicule the technology. Upon inspection of the array, it quickly became evident that there were no walkways in-between panels. Technicians informed me that they had been washing the array with a low-pressure hose but primarily expected them to be "cleaned by the rain." The rainy season lasts about 3 months of the year, and during the rest of the year there is no precipitation. The roof had not been designed to permit access to the modules.

Especially in the desert conditions of West Africa, strict adherence to a regular cleaning schedule is essential for optimal performance.

Rain will not wash off dirt, and especially bird soiling. In fact, rain can have a low pH and be acidic, which, over time can corrode the module frame and lead to leaks and breakages. As such it is preferable to use demineralized, or distilled, water with a pH of 7 for cleaning.

While not ideal, the technicians now use a pressure washer to clean the array.

The implementation of this simple, regular washing routine has increased the system's output back to within 10% of expected production.

Although O+M costs for PV are the lowest in the industry, regular cleaning is necessary for optimal production.

Conducting a cost-benefit analysis of washing a solar carport in Ouagadougou, Burkina Faso

In another desert West African country, Burkina Faso, technicians completed an assessment of lost output due to sand on their 380 kW solar carport. They cleaned one string after a 2-week period of not cleaning. The rest of the array was completely covered in a fine layer of sand. They determined that the clean string produced 5% more power than an unclean string.

Based on a cost-benefit analysis, in which they compared the cost of labor with the benefit of more output, the facility decided not to implement a cleaning schedule.

Clearly, determining the frequency of cleaning is site-specific. The author advised this facility to conduct another test after 1 to 3 months without cleaning because the reduction in output is compounded by the accumulation of multiple layers of sand and other particles.

In addition, as explained in this chapter, the visual inspection and thermographic scans required to detect and diagnose issues require a clean array.

Further, sand will increase the temperature of the modules by trapping heat, which, over time, can lead to decreased wire insulation resistance and even hot spots.

At this facility, the inverters were located outside in the hot and dusty atmosphere, regularly reaching air temperatures of up to 40°C +. While this was likely a design decision to minimize home run wire costs, the inverters will now require more frequent inspection and cleaning. In particular, the fan and exhaust will need to be checked and cleaned or the **mean time between failures** will decrease significantly.

Inverter FMEA part 1: diagnostic techniques and risk priority number

Potential failure mode/effect of failure/ cause of failure	*Severity*	*Probability*	*Detectability*	*Detectability/diagnostic technique*	*Risk priority number (RPN)*
Failure mode: DC ground fault *Effect*: The ground fault detection interrupter (GFDI) in grid-tie inverter trips and shuts OFF output or, if GFDI does not trip fire may result. *Cause*: Unintended electrical connections between live conductors and equipment grounding conductor (EGC) due to damaged conductor insulation, pinched wiring or incorrect electrical connections Note: A ground fault is the "undesirable condition of current flowing through the grounding conductor."[21] Ground fault current can be very high. In addition to being a safety hazard, this creates a fire hazard as bare metal is heated by the current flow.	**9** – Hazardous with warning	**5** – Moderate	**7** – Very low (if the inverter GFDI trips, finding the source of the ground fault requires thorough investigation; if the GFDI does not trip since it is larger than the ground fault current (common in large PV arrays), it can go undetected until a second fault and become a parallel path for high current)	Check inverter control panel for condition status *Fuse*: Check if inverters' GFDI is blown through **Continuity test**: 1 Verify circuit is de-energized by testing voltage. 2 Place metal caps of leads on metal ends of fuse. 3 If meter reading changes to low resistance fuse is OK. Fuse permits current flow. 4 If resistance reading is unchanged, resistance is 100% and fuse is blown. Replace fuse. *Conductor*: Apply a voltage on conductor and measure current to determine insulation resistance	**315**

Failure mode: Capacitor failure *Effect*: Inverter internal damage and shut off *Cause*: High temperatures evaporate electrolytes quickly in electrolytic capacitors, reducing life See Figure 3.30 for signs of capacitor failure	**9** – Hazardous with warning	**5** – Moderate	**5** – Moderate	Check inverter control panel for condition status	**225**
Failure mode: Over-temperature *Effect*: Inverter shut off *Cause*: When fan is not operating, caused by no power to fan; when fan is operating, caused by faulty sensor or clogged intake and exhaust filters	**6** – System inoperable with minor damage	**5** – Moderate	**5** – Moderate	Check inverter control panel for condition status Inspect the fans and air grills Run the "Fan-Test" function on inverter panel	**150**
Failure mode: Low power *Effect*: Underperformance *Cause 1*: Poor solar resource *Cause 2*: Shading or sensor *Cause 3*: Data acquisition system (DAS) error	**4** – System operable with significant degradation of performance	**7** – High	**5** – Moderate	Check inverter control panel for condition status	**140**

(*Continued*)

(Continued)

Potential failure mode/effect of failure/cause of failure	*Severity*	*Probability*	*Detectability*	*Detectability/diagnostic technique*	*Risk priority number (RPN)*
Failure mode: AC under or over voltage *Effect*: Inverter shut off *Cause*: Utility power out of range, and inverter cannot synchronize with the grid	**5** – System inoperable without damage	**5** – Moderate	**5** – Moderate	Check inverter control panel for condition status	**125**
Failure mode: Gating fault *Effect*: Inverter shut off *Cause*: Transistor failure See Figure 3.31 for consequences of a gating fault	**7** – System inoperable with equipment damage	**3** – Low	**5** – Moderate	Check inverter control panel for condition status Check insulated-gate bipolar transistor (IGBT) and inverter boards for discoloration (see Chapter 2) Measure IGBT temperature through IR imaging	**105**
Failure mode: DC under or over voltage *Effect*: Inverter will not start or shut off *Cause*: PV string voltage below minimum inverter input voltage or above maximum inverter input voltage	**5** – Low	**4** – Moderate	**5** – Moderate	Check inverter control panel for condition status Check that the module design followed correct string sizing method	**100**

Figure 3.30 Signs of capacitor failure.

Source: Max Page, PC Stats.[22]

Figure 3.31 Consequences of gating fault. Image shows a failed insulated-gate bipolar transistor (IGBT) that experienced a thermal runaway, burning the module.

Source: MD TAUSIF/Shutterstock.com.

Inverter FMEA part 2: corrective actions and new risk priority number

Potential failure mode	*Recommended action(s)*	*New severity*	*New probability*	*New detectability*	*New risk priority number (RPN)*
DC ground fault	For new installations, use a transformer-less inverter, which can sense ground fault current as low as 300 mA. This makes small **DC ground fault** currents detectable. • Identify source of DC ground fault. 1 For grid-tie inverters, one of the DC conductors is bonded to ground. If a fuse is blown (and has disconnected the circuit due to overcurrent), a ground fault is present. 2 Turn off the inverter and DC/AC disconnects, wait 5 minutes for capacitors to discharge, and remove and test GFDI continuity with ohmmeter. If there is excessive resistance, a fuse is blown. Do not replace fuse until fault is cleared. 3 To isolate inverter from PV array, remove positive and negative conductors from inverter. 4 Turn on DC disconnect to put live DC voltage on the conductors. 5 Measure voltage between the positive and negative conductors to obtain open circuit voltage of array. 6 Measure positive to ground, and negative to ground. There should be 0 volts to ground from either conductor if there is no ground fault. 7 If any voltage to ground is measured, locate the fault. Beginning at inverter, check each connection point all the way to PV array, including DC disconnect, and combiner box. 8 With the conductors in free air, check the open circuit voltage of each string (positive to negative), and positive to ground and negative to ground. Find the fault by determining which conductor has voltage to ground. 9 Test resistance on all conductors to ensure the fault is isolated to that one conductor. 10 Replace the wire(s). 11 Keep records of tests and replacements in case of future ground faults.[23,24]	**9** – Hazardous with warning	**4** – Moderate	**6** – Low	**216**

Capacitor failure	• Ensure inverter is located in shaded area below temperature maximum and fan is working. • Verify intake and exhaust filters are clear. • Replace capacitor.	**9** – Hazardous with warning	**3** – Low	**5** – Moderate	**135**
AC under or over voltage	• Confirm all breakers are on. • Check AC voltage and, if out of range, contact utility. If within range, perform manual restart. • If issues persist, keep track of incidents and present to utility.	**5** – Low	**4** – Moderate	**5** – Moderate	**100**
Low power	1 Calculate expected system output using data acquisition system or PVWatts. 2 Compare with readings of on-site performance meters. 3 If no difference, check that inverter output is operating in the MPPT range of the inverter (see Chapter 2). If it is, this indicates normal module degradation over time. 4 If modules are within 25-year warranty period, check for new shading conditions, such as vegetation growth and soiling. 5 Finally, run continuity tests on fuses, open circuit voltage and maximum power current tests on the strings, and take IR images of the cells to check for hot spots. 6 Remove any new shading. Repair/replace wiring, modules, and breakers as needed.	**4** – Very low	**5** – Moderate	**4** – Moderately high	**80**

(Continued)

(Continued)

Potential failure mode	*Recommended action(s)*	*New severity*	*New probability*	*New detectability*	*New risk priority number (RPN)*
Gating fault	• If under warranty contact manufacturer for repair/replacement of integrated circuit (IC). • If damage limited to IGBT, first replace driver circuit that controls the IGBT, and then the IGBT. • Test all capacitors and drivers for normal operation.[25]	**7** – System inoper-able with equipment damage	**2** – Low	**5** – Moderate	**70**
Over-tempera-ture	• If fan not operating – check power supply to fan. If good, replace fan. If bad, replace power supply. • If fan operating – check if sensor readings are accurate with thermal scan. If incor-rect, replace sensor. If not, inspect and clean intake and exhaust filters for excessive buildup. • Perform weekly inspection, and, as required, cleaning of fan, intake and exhaust fil-ters.	**6** – System inoper-able with minor damage	**3** – Low	**3** – High	**54**
DC under or over voltage	• DC under voltage indicates the string panel(s) maximum power voltage (Vmp) has decreased to under the minimum inverter input voltage due to high temperatures. Ventilate the modules or add module(s) to meet the minimum input voltage. • DC over voltage indicates the string panel(s) open circuit voltage (Voc) has increased to above the maximum inverter output voltage due to cold temperatures. Remove a module(s) or use an inverter with a higher maximum input voltage.	**5** – Low	**2** – Low	**5** – Moderate	**50**

Lithium ion[26] battery FMEA part 1: diagnostic techniques and risk priority number

Potential failure mode/effect of failure/cause of failure	*Severity*	*Probability*	*Detectability*	*Detectability/ diagnostic technique*	*Risk priority number (RPN)*
Failure mode: Lithium plating *Effect*: Excessive current causes lithium ions to accumulate on anode surface, known as lithium plating; this reduces the free ions, decreasing battery capacity and decreasing cycle life. *Cause*: Over voltage and low temperature	**9** – Hazardous with warning	**3** – Low	**5** – Moderate	Check battery control panel for status Check charge controller output voltage and current (see below)	**135**
Failure mode: Thermal runaway *Effect*: Fire *Cause*: Overheating at 70° C to 110° C from overcurrent, overcharging, internal or external short circuit or high ambient temperature breaks down organic solvents in electrolyte, releasing hydrocarbon gases, increasing pressure. At 200° C, cobalt cathodes decompose (other cathodes have a higher melting point), releasing oxygen that burns the electrolyte and gases, further increasing temperature and pressure until explosion. Cobalt breaks down at the lowest temperature, while iron phosphate ($FePO_4$) breaks down at the highest temperature. For this reason, $LiFePO_4$ is widely-considered the safest Li+ battery commercially available.	**9** – Hazardous with warning	**3** – Low	**4** – Moderately high	Check battery control panel for status Check charge controller output voltage and current	**108**

Figure 3.32 demonstrates how lithium ion battery performance varies along its state of charge. Optimal state of charge is 20% to 90%. Both undercharging and overcharging reduces battery life.

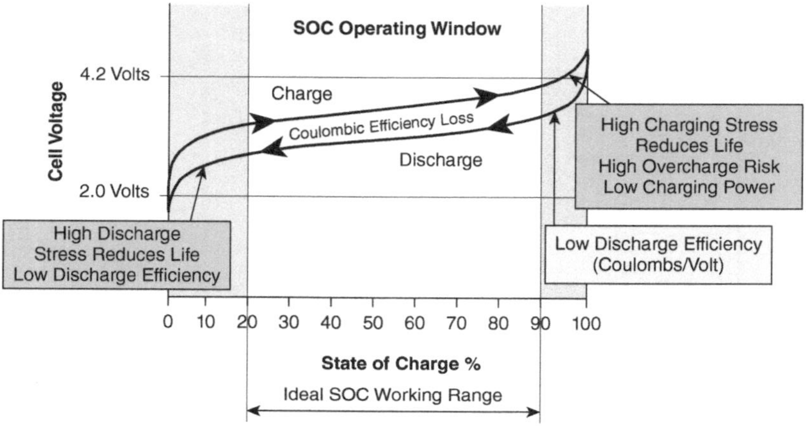

Figure 3.32 State of charge operating window.

Source: Barrie Lawson/Woodbank Communications Ltd.[27]

Figure 3.33 shows the effect of temperature on cycle life. The ideal working temperature range is 10° C to 60° C.

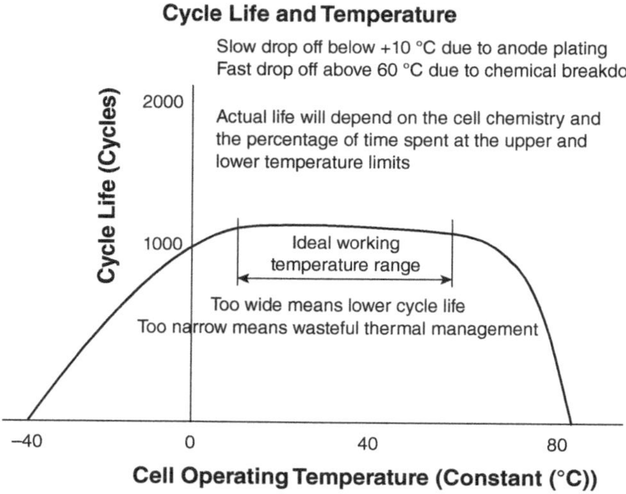

Figure 3.33 Lithium ion battery cycle life and temperature.

Source: Barrie Lawson/Woodbank Communications Ltd.[28]

Lithium ion battery FMEA part 2: corrective actions and new risk priority number

Potential failure mode	*Recommended action(s)*	*New severity*	*New probability*	*New detectability*	*New risk priority number (RPN)*
Lithium plating	• Store battery in shaded location where the temperature is within its operating range. • Prevent over voltage (above 4.2 volts per cell) and charging and discharging below 15° C. • Ensure charge controller is set to output voltage and current within acceptable parameters. • Ensure battery bank is monitored by Battery Management System (BMS) that shows and permits control of state of charge, current, voltage, and internal and ambient temperatures.	**9** – Hazardous with warning	**2** – Low	**3** – High	**54**
Thermal runaway	• Same as above, and: • Use a charge controller with maximum powerpoint tracking (MPPT), and temperature feedback to limit current if battery is overheating. • Use Li+ batteries with a combination of nickel manganese and cobalt as the cathode ($LiNiMnCoO_2$), often used for electric vehicles, or iron phosphate ($LiFePO_4$). The German battery firm Sonnen uses $LiFePO_4$ and, with thermal runaway at 518 F (270° C), advertises the greater safety and cycle life of their product. • Procure batteries with internal safety controls.	8 – Very high (system inoperable with destructive failure without compromising safety due to use of internal design features (thermal switch, separator shutdown, flame retardant, cell venting))	**2** – Low	**3** – High	**48**

To mitigate against risk, make sure to procure batteries with internal design controls, including:

- *Thermal switches* in circuitry that trip on excessive temperature.
- *Separator shutdowns* – use a polymer to separate the positive and negative electrodes in the battery. The polymer provides a path for the lithium ions to conduct and prevents direct contact between the electrodes. If the battery temperature near the polymer reaches melting point its pores will close, stopping further electrochemical reactions, and thus preventing thermal runaway.
- *Flame retardants (FRs)* – add FRs to the electrolyte to prevent electrolyte from releasing flammable gases and the generation of oxygen in the cathode.
- *Cell venting* – allows for a release of pressure due to gas accumulation to prevent further temperature rise.[29]

As shown in Figure 3.34, temperatures outside of the safe operating window, both high and low, have an adverse effect on lithium ion cells.

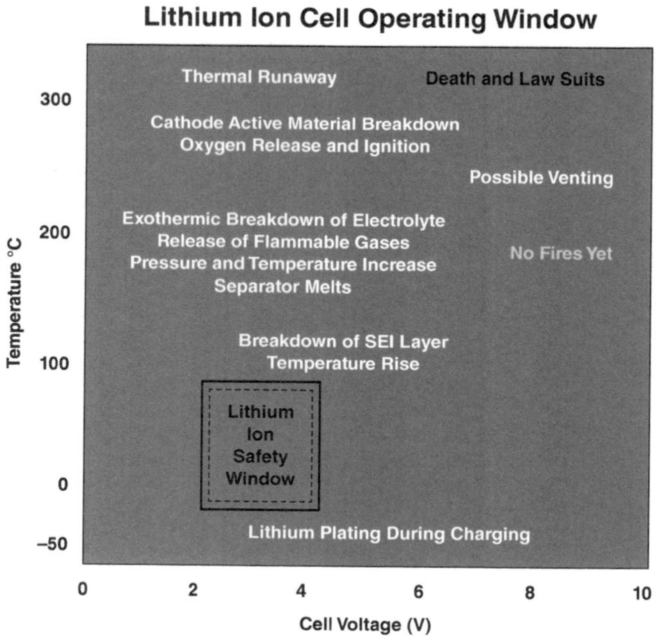

Figure 3.34 Lithium ion cell operating window.

Source: Barrie Lawson/Woodbank Communications Ltd.[30]

Other system components FMEA part 1: diagnostic techniques and risk priority number

Component/potential failure mode/effect of failure/cause of failure	*Severity*	*Probability*	*Detectability*	*Detectability/diagnostic technique*	*Risk priority number (RPN)*
Component: Charge controller *Failure mode*: Failure of internal component, e.g. capacitor, inductor, MOSFET, Schottky diode *Effect*: Controller off *Cause*: Overheating due to environmental conditions or loose connection	**9** – Hazardous with warning	**3** – Low	**4** – Moderately high	• Measure incoming voltage from PV array to charge controller. Measure voltage from charge controller to battery. • Measure current between charge controller and battery by connecting ammeter in series with charge controller and battery. Check your battery requirements – standard battery voltages are 12, 24, and 48 volts. Li+ batteries can be charged from 50 to 100% of their amp-hour rating, while lead acid batteries at 10% of their amp-hour rating. Overcharging gets released as heat that can damage the battery.	**108**
Component: Wiring *Failure mode*: Decreased insulation resistance *Effect*: Short circuits and ground faults, reduced production and potentially fire *Cause*: Wire sagging or pinching in hot climates causes thermal melting or mechanical separation; humidity and water infiltration leads to ground faults between live conductors and the EGC	7 – High	**5** – Moderate	**Low** – 6	• Measure insulation resistance (for 3-phase connections and inverters check the phase to phase and phase to ground insulation resistance) • Measure conductor path impedance • Thermographic inspection of terminals • Check EGC runs firmly through each module and to ground	**210**

(*Continued*)

(Continued)

Component/potential failure mode/effect of failure/cause of failure	*Severity*	*Probability*	*Detectability*	*Detectability/diagnostic technique*	*Risk priority number (RPN)*
Component: Racking *Failure mode*: Missing clips, bolts, WEEBs (washer electrical equipment bonding), rust, flashing not sealed *Effect*: Metals that are not bonded pose an electrical safety hazard and loose racking can cause injury to roofs and residents; unsealed flashing can cause water damage through roof openings *Cause*: Poor installation, severe weather conditions or, rarely, aging roof	**10** – Hazardous without warning	**2** – Low	**6** – Low	Proper installation and regular visual inspection throughout system life	**120**
Component: Circuit breaker/ground fault detection interrupter *Failure mode*: Failure of latching mechanism after repeated use *Effect*: Circuit permanently tripped until breaker replaced *Cause*: Repeatedly overloading circuit breaker current rating and/or current well above the breaker's nominal rating for continuous current	**5** – Low (system inoperable without damage, because circuit has tripped)	**4** – Moderate	**6** – Low	Check **time current curve** of circuit breaker and measure incoming current (see Figure 3.35). Repeated tripping usually indicates a short circuit or ground fault in your system that must be diagnosed through techniques explained above.	**120**

Component: Tracker *Failure mode*: Poor electrical connections, motor wear, controls error *Effect:* Loss of power, movement, and control *Cause*: Poor installation, motor bearing over or under greasing, sensor or software error	**4** – System operable with significant degradation of performance	**5** – Moderate	**5** – Moderate	• Visual inspection to check if sensor-reported status is accurate. • Check motor bearing.	**100**
Component: Tools *Failure mode*: Digital multimeter faulty readings or damaged internal components *Effect*: Incorrect diagnoses *Cause*: Lack of calibration, exceeding multimeter voltage rating, or exceeding current blows out fuse See Figure 3.36 for multimeter voltage categories	**7** – System inoperable with equipment damage	**7** – High	**6** – Low	• Calibrate digital multimeter (see the “Meters” section below). • Use the appropriate CAT and do not exceed voltage limits of multimeter.	**294**

(*Continued*)

(Continued)

Component/potential failure mode/effect of failure/cause of failure	*Severity*	*Probability*	*Detectability*	*Detectability/diagnostic technique*	*Risk priority number (RPN)*
Component: Transformer (step up, oil-cooled, distribution or substation) *Failure mode*: Contaminated insulating oil, primary or secondary windings, decreased insulation resistance *Effect*: Power loss *Cause*: For large PV systems, transformers are needed to increase the output voltage from the inverter (and at 3 phase) for interconnection with the grid. Transformer failure will prevent interconnection and prevent export and sale of energy from the PV system	**7** – System inoperable with equipment damage	**4** – Moderate	**5** – Moderate	• Insulation resistance test. • Insulation power factor (doble) test. • Oil analysis. • Ratio and phase relation test. • NETA mechanical/visual and electrical tests. • Check phase balance. For procedures see *ANSI/IEEE C57.12.90–1993, IEEE Standard Test Code for Liquid-* *Immersed Distribution, Power and Regulating Transformers and IEEE Guide for Short Circuit* *Testing of Distribution and PowerTransformers.*[31]	**140**

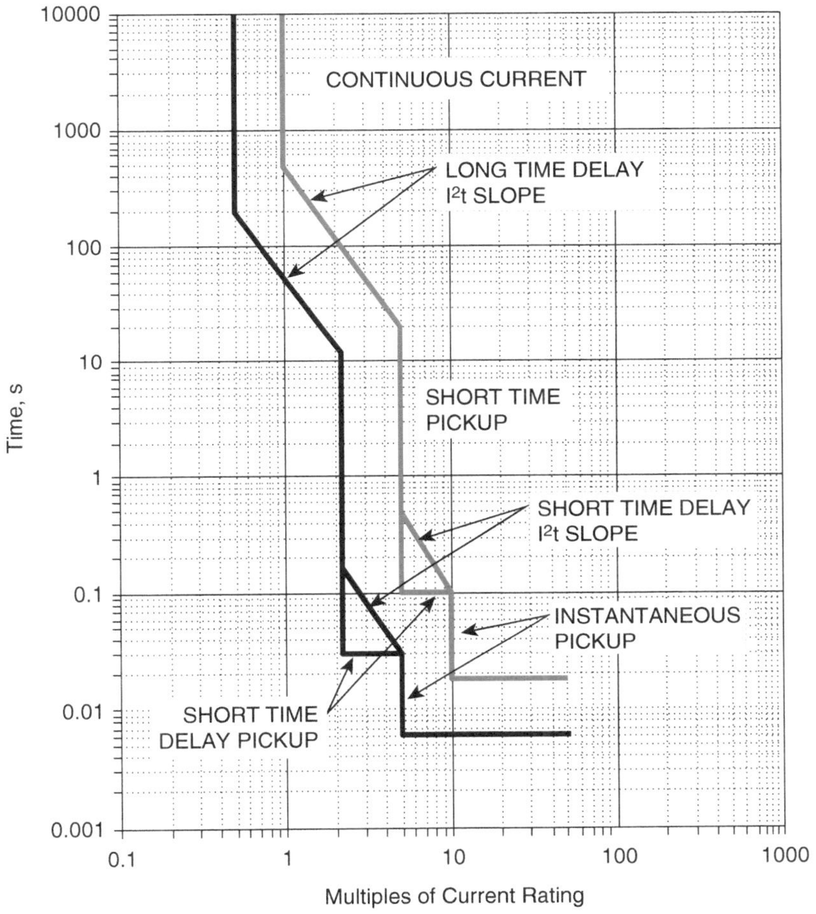

Figure 3.35 Time current curve of a circuit breaker.

Source: Ver Pangonilo.[32]

As shown in Figure 3.35, the TCC curve shows that circuit breaker trip speed increases with current. Circuit breakers can be programmed based on how long a certain current above nominal should be permitted. For instance, it is necessary to exceed the nominal continuous current at times when certain equipment, such as motors, are started to allow for the brief inrush current needed to start the machine. Repeated tripping will cause the latching mechanism to fail and indicates either a short circuit, such as a ground fault, or an improperly designed electrical system.

Diagnosing circuit breakers requires understanding how they operate under varying amounts of current. The time current curve (TCC), typically displayed on a circuit breaker, shows how quickly the breaker will trip based on the current to which it is exposed.

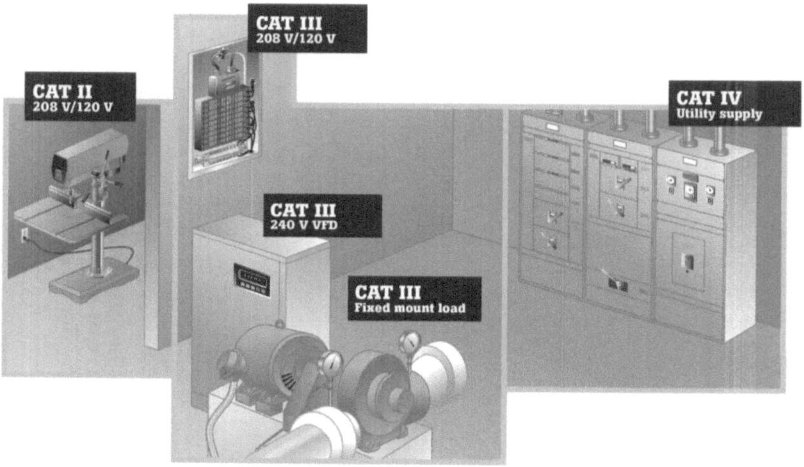

Figure 3.36 Multimeter voltage categories. A higher category means great ability to withstand voltage spikes and transients. CAT I is for consumer electronics, CAT II for appliances, CAT III for switchgear, and CAT IV for utility connections and high altitudes.

Source: Reproduced with permission, Fluke Corporation.[33]

Other system components FMEA part 2: corrective actions and new risk priority number

Potential failure mode	*Recommended action(s)*	*New severity*	*New probability*	*New detectability*	*New risk priority number (RPN)*
Charge Controller Failure of internal component	• Depending on extent of internal damage, replace part(s) or procure new charge controller. • Ensure charge controller in shaded area where ambient temperatures are within device operating range. • Ensure charge controller connected to data acquisition system, which can be conveniently and remotely accessed by technicians.	**9** – Hazardous with warning	**2** – Low	**2** – Very high	**36**
Wiring Decreased insulation resistance	• Replace melted or damaged insulation and conductors. Test new insulation to ensure high resistance and conductor(s) to ensure low impedance. • For insulation, ensure insulation resistance equals or is greater than manufacturer specified minimum resistance. • Use insulation with a higher insulation material constant (K value). See Insulated Power Cable Engineers Association (IPCEA) for details on testing procedures.[34] • Perform regular IR imaging of wiring and correct wire sagging.	**6** – System operable with significant degradation of performance (assuming insulation material with higher K value used)	**2** – Low	**4** – Moderately high	**48**
Racking Missing clips, bolts, WEEBs (washer electrical equipment bonding), rust, flashing not sealed	• Attach missing components and fix flashing. • Apply an anti-corrosive coating to rusting metal racking or use racking with high resistance to corrosion. • Perform regular visual, and, if large system, aerial inspections. • Replace roof shingles if necessary.	**9** – Hazardous with warning	**1** – Remote	**4** – Moderately high	**36**

Circuit Breaker Latching mechanism failure; ground fault detection interrupter (GFDI) fuse trip	• Check system for ground fault or other short circuit and correct. Only then, replace GFDI fuse or circuit breaker, ensuring an appropriate current rating and the time current curve is programmed properly for your electrical system. • Ensure circuit breaker status connected to building monitoring system.	**5** – Low	**3** – Low	**5** – Moderate	**75**
Tools digital multimeter faulty readings	• Use multimeter rated for required CAT level and voltage category. Do not exceed. • Use well-insulated multimeters. • Calibrate multimeter before use. • Be sure to test DC voltage on DC setting and AC voltage on AC setting. • Replace blown out fuse with same current rating. • Replace multimeters after frequent use, per manufacturer's schedule.	**7**	**3** – Low	**2** – Very High	**42**
Transformer	• Replace contaminated oil. • Perform IR imaging of insulation and coils and replace any damaged wiring and insulation. • Check, and if needed, replace tap changer (for changing transformer output voltage) and bushings (used to dampen energy, reducing vibrations). • Ensure power factor does not exceed 0.5% at 20 °C (per ASTM D24).[35] • Regularly inspect transformer tank to ensure no oil leakage and test oil quality. • Ensure emulsified water (water suspended in oil) does not exceed 25 ppm at 20 C. (per ASTM D 95–13[36]). • Ensure no more than 0.03 grams potassium hydroxide (KOH) per ml (per ASTM D 974).[37] • Ensure transformer connected to building monitoring system.	**6** – System inoperable with minor damage	**3** – Low	**3** – High	**54**

Meters

Conductor insulation should be tested to assess damage and locate faults within an array. Over time wiring insulation degrades, causing a decrease in resistance. This allows a small current to flow through the insulator, resulting in system energy loss. Insulation resistance is tested with an insulation tester, or megohmeter, in the range of Mohms (megohms) by applying a test voltage on a conductor and measuring resistance between the conductor and ground, or other conductors. Conductor insulation resistance should be high, and grounding electrode resistance low. When testing conductor insulation or circuits with electronics sensitive to the high test voltages used, circuits must be discharged before and after tests. Figure 3.37 shows a sample insulation tester.

Figure 3.37 Insulation tester (meghohmeter).

Source: Reproduced with permission, Fluke Corporation.

The red probe of the digital multimeter shown in Figure 3.38 is connected to the red port (VΩ) and its tip to the positive terminal of an AC circuit (as the voltage is set to AC). The black probe is connected to the black port (COM, for Common) and its tip to the negative terminal of an AC circuit.[38] See Figure 3.39 for a technician reading the DC voltage of a battery using a multimeter. Be sure that the instrument is set to measure AC voltage when testing AC circuits and to DC voltage when testing DC circuits. Be sure that the current settings are set to the type of electricity you are measuring as well, although some multimeters have both on the same setting. If unsure of the magnitude of the voltage or current, set the knob to the highest setting and work down. Do not exceed the maximum current rating of the multimeter or the fuse will blow.

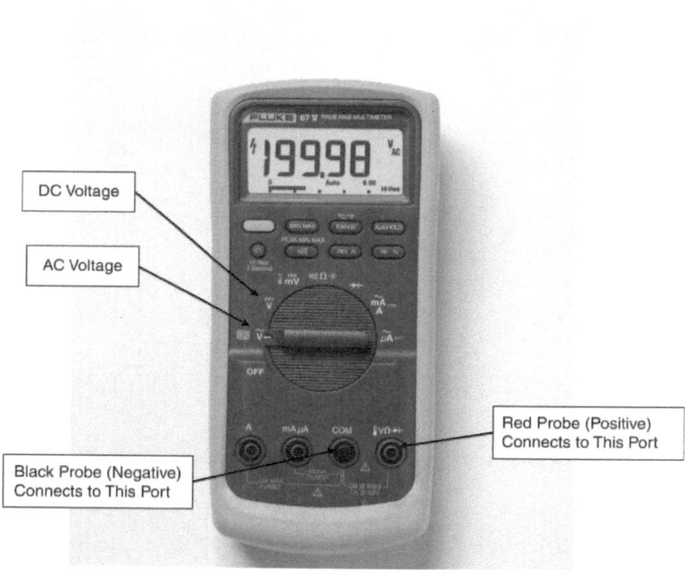

Figure 3.38 Digital multimeter.

Source: Reproduced with permission, Fluke Corporation.

Digital multimeter calibration procedure

1. Set multimeter to highest resistance range (highest ohm setting).
2. Touch the test probes together (display should show '0' ohms).
3. If display does not read '0,' press calibration button until display shows '0.'

Figure 3.40 shows a technician using a clamp-on meter to read the induced current from a wire carrying AC power.

Figure 3.39 Reading battery DC voltage using a multimeter. The right terminal of a battery is positive and connected to the red probe, and the left terminal of a battery is negative and connected to the black probe. The selection knob is set to DC voltage up to 20 volts. Technician is wearing non-conductive safety gloves.

Source: Pryzmat/Shutterstock.com.

Figure 3.40 Clamp-on meter – induced current read off a wire.

Source: Reproduced with permission, Fluke Corporation.

Maintenance schedules

Based on the FMEA assessment, your facility is now ready to develop a maintenance schedule. The sample schedule in Table 3.4 should be tailored to the features of your site. In particular, consider your climate, battery technology, and the economics of electricity, both purchased and exported, in your area.

Table 3.4 Sample solar PV system maintenance schedule

Maintenance task (responsible individual(s))	*Daily*	*Weekly*	*Monthly*	*3 months*	*Annual*	*Biennial (once every 2 years)*
Complete system visual inspection		X				
Array washing		X				
Battery cleaning			X			
Battery "top-up" (while lead acid batteries are discouraged, if used it will be necessary to ensure they are at 100% state of charge)			X			
Inverter cleaning			X			
Charge controller cleaning			X			
Aerial inspection and infrared imaging of modules, inverters, charge controllers, and batteries				X		
Wiring insulation testing					X	
Transformer testing						X

Templates

See Figure 3.20 for legends of severity, probability, and detectability.

FMEA part 1: diagnostic techniques and risk priority number

Potential failure mode/effect of failure/cause of failure	*Severity*	*Probability*	*Detectability*	*Detectability/diagnostic technique*	*Risk priority number (RPN)*

FMEA part 2: corrective actions and new risk priority number

Potential failure mode	*Recommended action(s)*	*New severity*	*New probability*	*New detectability*	*New risk priority number (RPN)*

Solar PV system maintenance schedule

Maintenance task (responsible individual(s))	*Daily*	*Weekly*	*Monthly*	*3 months*	*Annual*	*Biennial (once every 2 years)*

Notes

1 Commonly Used Statistics. *OSHA.gov*. Accessed January 1, 2019. www.osha.gov/oshstats/commonstats.html.

2 TABLE A-9. Fatal occupational injuries by event or exposure for all fatal injuries and major private industry(1)sector, all United States, 2017. Accessed January 1, 2019. www.bls.gov/iif/oshwc/cfoi/cftb0321.htm.

3 *Solar Photovoltaic (PV) System Safety and Fire Ground Procedures*. Publication. San Francisco Fire Department. San Francisco, CA, 94107: San Francisco Fire Department, 2012. http://ufsw.org/pdfs/photovoltaic_systems.pdf.

4 See National Electrical Code Article 250 for grounding and bonding requirements for specific installations.

5 These requirements are current per the NEC 2017 and effective as of 2019.

6 FallProof Systems LLC. www.fallproof.com/guardrails-and-skylights/rooftop-railings-and-walkways/.

7 Hard Hats. https://us.msasafety.com/Head-Protection/Hard-Hats/c/11204.

8 Wind-Shield, A.F. Global Vision Eyewear. www.globalvisioneyewear.com/product/wind-shield-af/.

9 Parrish, P. *Photovoltaic Laboratory: Safety, Code-Compliance, and Commercial Off-the-Shelf Equipment*. 1st Edition. CRC Press. Boca Raton, FL. April 7, 2016.

10 "Personal Fall Arrest System." *Weeklysafety.com*. https://blog.weeklysafety.com/blog/safety-topics/fall-arrest-systems.

11 "Suspension Trauma Safety Straps." 3M. www.3m.com/3M/en_US/company-us/all-3m-products/~/3M-DBI-SALA-Suspension-Trauma-Safety-Straps-9501403-1-EA/?N=5002385+3291827097&rt=rud.

12 "Balance-of-System Equipment Required for Renewable Energy Systems." US Department of Energy.

13 *Best Practices in Photovoltaic System Operations and Maintenance*. 2nd Edition. NREL/Sandia/Sunspec Alliance SuNLaMP PV O&M Working Group. www.nrel.gov/docs/fy17osti/67553.pdf. December 2016.

14 Ibid.

15 Jordan, Dirk C. and Sarah R. Kurtz. Photovoltaic Degradation Rates – An Analytical Review. Report no. NREL/JA-5200-51664. Golden, CO: National Renewable Energy Laboratory, 2012. www.nrel.gov/docs/fy12osti/51664.pdf.

16 Photovoltaic Degradation Rates – An Analytical Review, D.C. Jordan and S.R. Kurtz. NREL. www.nrel.gov/docs/fy12osti/51664.pdf. June 2012.

17 Haney, J. and A. Burstein. *PV System Operations and Maintenance Fundamentals*. US Department of Energy. www.solarabcs.org/about/publications/reports/operations-maintenance/pdfs/SolarABCs-35-2013.pdf. August 2013.

18 Chen, J. "Evaluating thermal imaging for identification and characterization of solar cell defects." Iowa State University. https://lib.dr.iastate.edu/cgi/viewcontent.cgi?article=4980&context=etd. 2014.

19 "Cell Guard™ Solar Cell Encapsulation for Making Solar Panels." MLA Solar. www.mlsolar.com/cell-guard-solar-cell-encapsulation-for-making-solar-panels/.

20 National Renewable Energy Laboratory, Sunspec Alliance SuNLaMP, and O&M Working Group. *Best Practices in Photovoltaic System Operations and Maintenance*, 2nd Edition. Technical paper no. NREL/TP-7A40-67553. National Renewable Energy Laboratory. Golden, CO: National Renewable Energy Laboratory, 2016. www.nrel.gov/docs/fy17osti/67553.pdf.

21 Dunlop, J.P. *Photovoltaic Systems*, 2nd Edition. National Joint Apprenticeship and Training Committee (NJATC). July 1, 2009.

22 "Blown, Burst and Leaking Motherboard Capacitors – A Serious Problem?" Max Page, PC Stats. www.pcstats.com/articleview.cfm?articleID=195.

23 "Suggestions on How to Troubleshoot a Ground Fault." Global Specialist in Energy Management and Automation. Accessed March 24, 2019. www.schneider-electric.us/en/faqs/FA304871/.

24 Technology, SMA Solar. "Tech Tip: Troubleshooting a Ground Fault." YouTube. April 10, 2013. Accessed March 24, 2019. www.youtube.com/watch?v=GEyP5dUCvdw.

25 "How to Troubleshoot and Repair Any Variable Frequency Drive." How to Repair Any VFD | Details | Hackaday.io. Accessed March 24, 2019. https://hackaday.io/project/24909-how-to-repair-any-vfd/details.

26 As explained in Chapter 2, it is not advised to use lead acid batteries for PV systems. The FMEA is thus focused on lithium ion batteries.

27 Lawson, B. "State of Charge Operating Window." Woodbank Communications Ltd. www.mpoweruk.com/lithium_failures.htm.

28 Lawson, B. "Lithium Ion Battery Cycle Life and Temperature." Woodbank Communications Ltd. www.mpoweruk.com/lithium_failures.htm.

29 Kong, L., C. Li, J. Jiang, and M. Pecht. "Li-Ion Battery Fire Hazards and Safety Strategies." *Energies* 11, no. 9 (2018): 2191. doi: 10.3390/en11092191.

30 Lawson, B. "Lithium Ion Cell Operating Window." Woodbank Communications Ltd. www.mpoweruk.com/lithium_failures.htm.

31 C57.12.90-1987 – IEEE Standard Test Code for Liquid-Immersed Distribution, Power, and Regulating Transformers and IEEE Guide for Short-Circuit Testing of Distribution and Power Transformers. IEEE. https://ieeexplore.ieee.org/document/35049. 1988.

32 Pangonilo, V. "Circuit Breaker Time-Current Characteristic Curve Made Easy." Accessed September 18, 2009. https://pangonilo.com/2009/09/circuit-breaker-time-current-characteristic-curve-made-easy.html.

33 "The Danger of Transients." Fluke. www.fluke.com/en-us/learn/best-practices/safety/test-tool-safety/the-danger-of-transients.

34 "A Stitch in Time: The Complete Guide to Electrical Insulation Testing." 2006. Accessed March 24, 2019. www.biddlemegger.com/biddle/Stitch-new.pdf.

35 Standard Test Method for Acid and Base Number by Color-Indicator Titration. ASTM D974. Standard Test Method for Dissipation Factor (or Power Factor) and Relative Permittivity (Dielectric Constant) of Electrical Insulating Liquids. ASTM D924-15. www.astm.org/Standards/D924.htm.

36 ASTM D95-13. Standard Test Method for Water in Petroleum Products and Bituminous Materials by Distillation. www.astm.org/Standards/D95.htm.

37 www.astm.org/Standards/D974.htm.

38 While these can be reversed for an AC circuit, for a DC circuit, the red probe must go to the positive terminal, and the black probe must go to the negative terminal. If they are reversed in a DC circuit the polarity will be reversed and the reading will be negative.

4 Tools and methodology for monitoring, measurement, and verification

Monitoring

The O+M discussed in Chapter 3 relies on power monitoring. It is essential to a well-functioning PV system. By monitoring the functioning of the various components, we can detect existing or imminent issues in order to keep the system operating at an optimal or, at least satisfactory, level. This monitoring can give early notice of when maintenance is warranted and thus avoid unexpected downtime. In addition, the information gathered about energy generation and usage helps to increase the efficiency and power generation of the system, which improves financial results.

A power monitoring system is composed of three parts: first, a *metering or monitoring device* or devices to report on power generated and used; second, *software* to record and analyze the information produced; and third, a *communications link* between the metering devices and the software package. The quality and capabilities of each component can range from simple to very high grade depending on the needs of the customer.[1]

It is essential that the power monitoring system come with a display, known as a **graphical user interface (GUI)** or **human machine interface (HMI)**. In today's day and age, with the widespread proliferation of information communication technologies, the need for an HMI goes without saying, but in the early days of power monitoring systems this was not always the case.

The Energy Policy Act of 2005 authorized the U.S. Federal Energy Management Program (FEMP), a part of the U.S. Department of Energy (DOE), to mandate the use of **energy data systems (EDS)** for federal buildings over 5,000 ft^2 (465 m^2). EDS simply collect data from sensors without displaying it. A facility team can do nothing with the data if they cannot monitor it. Now, facilities that had EDS are upgrading them to **energy information systems (EIS)** so the data can be viewed and analyzed. To display the data a facility will often use a web-based system with a *Transmission Control Protocol/Internet Protocol (TCP/IP)*. The many meters recording the data from your system send data by TCP/IP over *Local Area Networks (LAN)*, typically

Ethernet, but increasingly Wi-Fi. Figure 4.1 shows the GUI of a solar energy system including the daily and historical production and earnings.

With the rise in popularity of module level power electronics (MLPEs), such as microinverters, HMIs can now display the conditions of each solar module in real time, as shown in Figure 4.2.

There is a wide and growing range of power monitoring systems. Some of the better-known systems include Tesla's power monitoring system, Enphase's Envoy, APsystems' Energy Monitoring & Analysis, and Solar-Log.

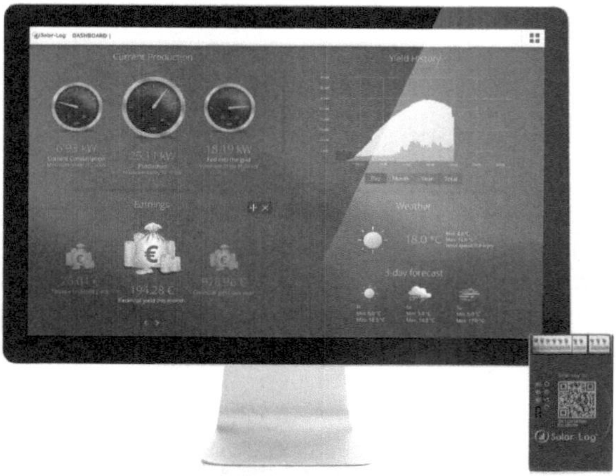

Figure 4.1 Solar-Log dashboard.

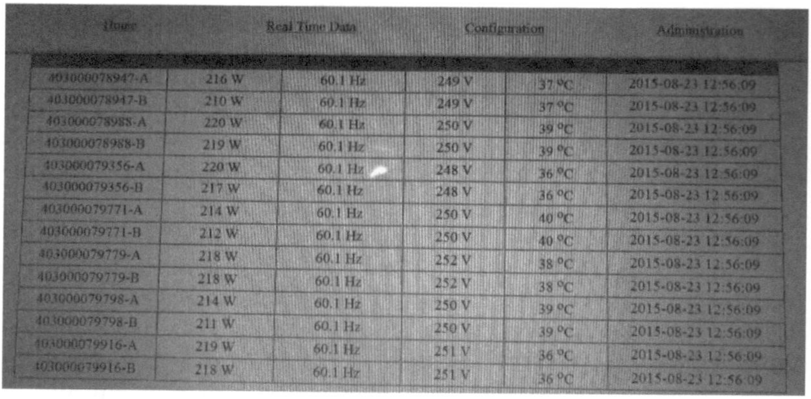

Home · Real Time Data · Configuration · Administration

403000078947-A	216 W	60.1 Hz	249 V	37 °C	2015-08-23 12:56:09
403000078947-B	210 W	60.1 Hz	249 V	37 °C	2015-08-23 12:56:09
403000078988-A	220 W	60.1 Hz	250 V	39 °C	2015-08-23 12:56:09
403000078988-B	219 W	60.1 Hz	250 V	39 °C	2015-08-23 12:56:09
403000079356-A	220 W	60.1 Hz	248 V	36 °C	2015-08-23 12:56:09
403000079356-B	217 W	60.1 Hz	248 V	36 °C	2015-08-23 12:56:09
403000079771-A	214 W	60.1 Hz	250 V	40 °C	2015-08-23 12:56:09
403000079771-B	212 W	60.1 Hz	250 V	40 °C	2015-08-23 12:56:09
403000079779-A	218 W	60.1 Hz	252 V	38 °C	2015-08-23 12:56:09
403000079779-B	218 W	60.1 Hz	252 V	38 °C	2015-08-23 12:56:09
403000079798-A	214 W	60.1 Hz	250 V	39 °C	2015-08-23 12:56:09
403000079798-B	211 W	60.1 Hz	250 V	39 °C	2015-08-23 12:56:09
403000079916-A	219 W	60.1 Hz	251 V	36 °C	2015-08-23 12:56:09
403000079916-B	218 W	60.1 Hz	251 V	36 °C	2015-08-23 12:56:09

Figure 4.2 Module operating data displayed from Microinverter APsystems Energy Monitoring & Analysis.

Source: Used with permission. NABCEP® is a registered trademark owned by the North American Board of Certified Energy Practitioners® (NABCEP®).

Tesla's power monitoring system allows the user to track the solar energy produced by a PV system, the power stored and drawn from its Powerwall battery, and home energy use. It operates in real time and helps to reduce the owner's reliance on utility grid power. The software lets the user see daily, weekly, monthly and annual energy generation and consumption, and displays a history of utility power outages. In addition, the system can be customized to the consumer's need, including regulating when to draw on the energy stored in the Powerwall, such as at night, or only in case of a power outage.[2]

Enphase's Envoy is a communications gateway that gathers information from metering devices on microinverters and transmits it to a software system for recording and analysis. This system uses the home's electrical wiring, rather than Wi-Fi, for its communications. From there, the information and analysis are available to the user by Wi-Fi or a "mini-LAN."[3] Figure 4.3 shows the GUI for Enphase's Envoy.

APsystems' Energy Monitoring & Analysis (EMA) is a package of monitors and software that provides the user with real-time web-based monitoring for analysis and control of each individual module and microinverter in the solar array, thereby optimizing performance over the system's lifetime. The EMA uses an Energy Communication Unit as the information gateway to microinverters connected to the panels.[4]

Solar-Log provides a power monitoring system for solar panels with residences and business and for PV plants. It can warn when a malfunction occurs, such as the failure of an inverter, the partial failure of a module, or cable damage. In this way, yield losses can be minimized due to early detection. The information and analysis are available through the Internet via a special portal.[5]

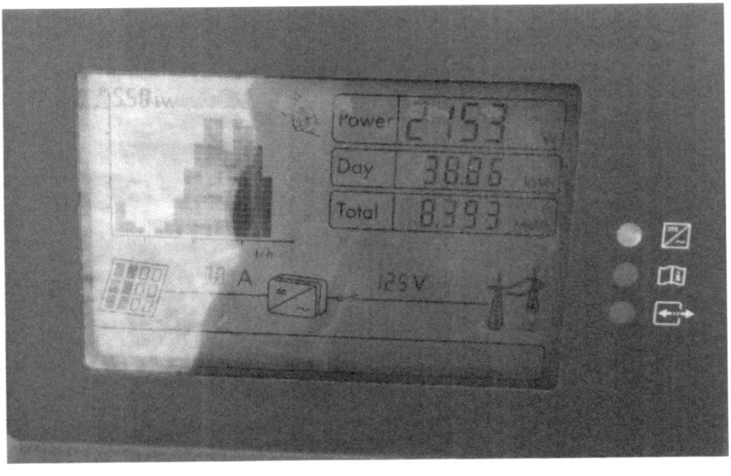

Figure 4.3 Enphase microinverter performance analytics.

Source: NABCEP® PV Study Guide. Used with permission. NABCEP® is a registered trademark owned by the North American Board of Certified Energy Practitioners® (NABCEP®).

The frontier of building controls and data analysis

Taking it one step further, a facility can have an **energy management system (EMS)** that allows the user to control the system. EMS were first developed for use by large facilities that sought to optimize and economize their heating ventilation and air conditioning (HVAC) systems. Today an EMS for a PV system can be used to correct back to *setpoints*, desired operating settings, for all system components. When conditions change, the control system uses negative feedback to correct the error back to the setpoint. In particular, PV systems with trackers can benefit from the use of a **programmable logic controller (PLC)** that corrects errors through a response that is proportional to the magnitude of the error. These most-basic PLCs operate by **proportional gain**, which is a response proportional to the size of the difference between the setpoint and the current value (or control point) multiplied by an adjustable constant. As shown in Figure 4.4, the greater the constant, the quicker the response, but this leads to *overshoot*, an overcorrection beyond the setpoint.

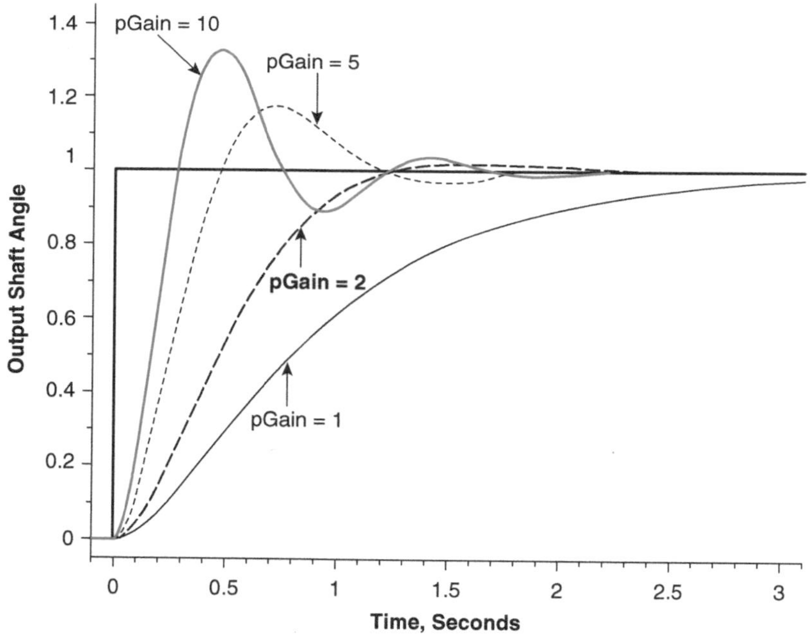

Figure 4.4 Motor and gear with proportional gain over time. The stationary line at 1 is the setpoint. The optimal setting for this device with a proportional gain (pGain) of 2 is shown in bold.

Source: Tim Wescott, "PID Without a PhD." http://wescottdesign.com/articles/pid/pidWithoutAPhd.pdf.

The next level of control added to proportional is called **integral gain**, in which the controller increases the movement towards the setpoint by factoring in the duration of past errors. See Figure 4.5.

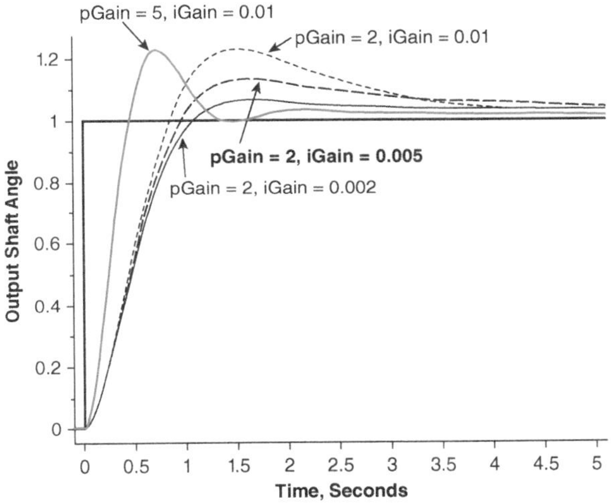

Figure 4.5 Motor and gear with proportional and integral (PI) gain over time. Compared to only proportional gain, PI more quickly returns to the setpoint although with more overshoot. The stationary line at 1 is the setpoint. The optimal setting is shown in bold (pGain = 2, iGain = 0.005).

Source: Tim Wescott, "PID Without a PhD." http://wescottdesign.com/articles/pid/pidWithoutAPhd.pdf.

Finally, with **derivative gain**, the controller accounts for the rate of change of the error, and the setpoint is achieved with less overshoot and oscillations. The proportional gain corrects based on present behavior, the integral gain on past behavior, and the derivative gain on future predictions. As shown in Figure 4.6, a controller that uses all three parameters is known as a **proportional, integral, derivative gain (PID) controller**.

While most powerful, the PID controller is also more prone to issues due to greater chance of noise based on differences in sampling times. This is why PI gain controllers are the most widely used by industry.

When PV systems are connected with other energy sources, such as wind turbines, combined cycle power plants, and diesel generators in a **microgrid**, an EMS is essential for effective system operation.

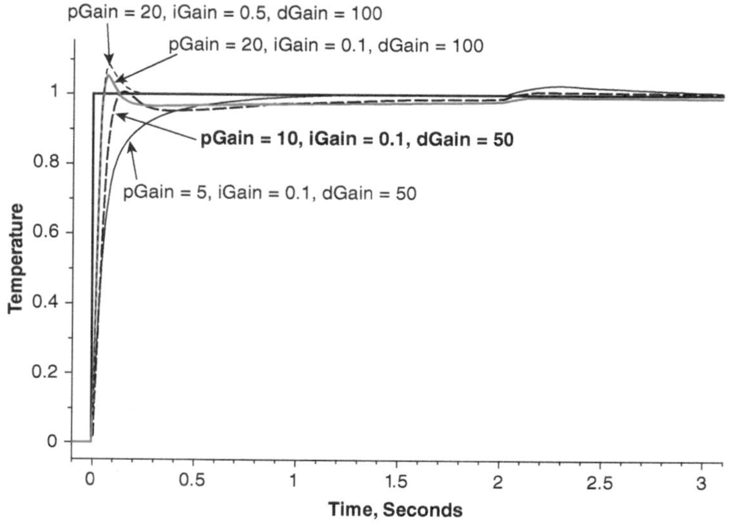

Figure 4.6 Heater with proportional, integral, and derivative (PID) gain over time. Compared to PI, PID most quickly returns to the setpoint and with minimal overshoot. The stationary line at 1 is the setpoint. The optimal setting is shown in bold (pGain = 10, iGain = 0.1, dGain = 50).

Source: Tim Wescott, "PID Without a PhD." http://wescottdesign.com/articles/pid/pidWithoutAPhd.pdf.

Machine learning

This takes us to the cutting edge of controls and data analytics. The use of **machine learning (ML)**, an algorithm that learns or draws patterns from data, is transforming almost every industry today. The optimal, or at a minimum efficient, operation of a solar energy system is complicated by a variety of factors that are not present with a fossil fuel plant. Varying cloud coverage and dust levels affect power generation, compounding the chance of problems. ML can enable us to better understand these risks and empower us to take steps to improve efficiency and address potential problems. Fortunately, a system or facility will often have a multitude of sensors to report on current conditions. The challenge becomes how to analyze the data to improve O+M. Gems can be hidden within the mountain of data generated, if only they can be found.

With the computing power readily available today, it is possible to sift through the mounds of data to find valuable patterns and make useful predictions. Data analytics techniques exist that can perform this analysis and now can do what comes naturally to humans – learn from experience. ML is invaluable for processing massive amounts of data.[6] ML uses one of two techniques – **supervised learning** or **unsupervised learning**.

Supervised learning uses classification and regression to develop predictive models based on the evidence (data). Known information upon which future predictions are made is called **training data**. The algorithm first learns from the training data and then, when new data arrives, automatically classifies it into one of the pre-determined categories. In regression, the training data has both an input variable (x) and an associated output variable (y). When the new input data (x) is fed into the algorithm, it can then predict an output variable (y) for that data based on the training set.

On the other hand, in unsupervised learning, on the other hand, the algorithm has no prior known information, and thus there is no training data. Rather, it identifies patterns within the data. Cluster analysis, in which data is grouped based on similarities, is one of the most common techniques and is often used for image analysis and pattern recognition.[7]

As the solar energy sector matures, there is growing experience with the use of ML. For example, linear regression, a form of supervised learning, allows us to predict the 6-hour ahead energy that will be generated by a solar array.[8,9]

Machine learning models can allow a solar farm to perform predictive maintenance. This analysis can predict the expected, or mean, time between failures for key components, so that they can be repaired or replaced before they fail.[10]

Besides improving maintenance and preventing equipment failure, machine learning can enhance the efficiency of a solar energy facility. NEXTracker, which provides solar tracking equipment, reported that it saw gains of between 2% and 6% per year with this technique. ML can forecast cloud cover, fog, and dust, and make adjustments to solar PV panel arrays.[11] On a larger scale, Con Edison partnered with Columbia University to apply ML techniques to the New York City power grid. The results of the analysis of decades of data led to predictive maintenance on key transmission components. Con Edison was able to proactively address likely equipment failures before they occurred.[12]

ML offers powerful informational and predictive abilities that can improve maintenance and help guide proactive predictive maintenance, enhancing the performance of solar PV projects and facilities.

Measurement and verification

The return on investment (ROI) of a PV system can only be measured and verified through data. Standardization of M+V processes is needed to instill confidence in stakeholders, particularly so investors can be confident that their expected returns are backed up by actual data.

Developed over the last two decades by industry and government experts, the **International Performance Measurement and Verification Protocol (IPMVP)** is the world's leading procedure for measurement and verification to ensure accurate assessment of performance and financial metrics for energy efficiency and renewable energy projects. The IPMVP is published and updated by the Efficiency Valuation Organization (EVO), a nonprofit organization, with the advice of international experts.[13] The IPMVP builds on work begun in the 1990s by the U.S. Department of Energy, with industry and government experts from North America.

The EVO highlights closely-related reference publications that can assist in understanding and applying the concepts in the IPMVP. In particular, it points to the DOE's guidance for measurement and verification – *M&V Guidelines: Measurement & Verification for Performance Based Contracts, Version 4.0, Federal Energy Management Program (FEMP)*.[14] While the M+V guidelines focus on energy efficiency and water efficiency projects, they are helpful for renewable energy projects in terms of monitoring and verifying cost savings.

The IPMVP stresses the importance of preparing an M+V plan to properly determine and verify savings from a project. Advanced planning in developing a project will ensure that the data needed to make a proper savings determination will be available after implementation of the project. Further, once a savings report has been prepared, a third party may be brought in to verify that it complies with the M+V plan. The third party could also verify that the M+V plan itself is consistent with the objectives of the project.

Overall, the IPMVP defines broadly accepted M+V options and terminology for measuring savings, and provides a framework for developing a M+V plan. The IPMVP calls for cost savings reports to be prepared routinely in order to properly manage results and cash flow. These savings reports should follow the project's M+V plan and, in particular, should contain:

- Raw data for the reporting period (energy and independent variables).
- Description of any facility changes warranting adjustments to the baseline, and calculation of the necessary adjustment.
- Energy price used.
- Computed savings in energy and monetary units.[15]

The IPMVP is not meant to prescribe contractual terms, but rather to provide helpful and standard guidance on some key issues. The IPMVP can help in the selection of the M+V approach that best matches i) project costs and savings magnitude, ii) technology-specific requirements, and iii) risk allocation between buyer and seller, i.e. which party is responsible for installed equipment performance and which party is responsible for achieving long-term energy cost savings. The use of well-accepted definitions and practices set forth in the IPMVP facilitates financing by providing investors with confidence.[16]

Another valuable tool for designing and verifying systems is the DOE National Renewable Energy Laboratory's (NREL) **System Advisor Model (SAM)**. It is freely available to the public and calculates he performance and financial metrics of renewable energy projects. See https://sam.nrel.gov/. SAM is designed to help project developers and others, such as policy makers and equipment manufacturers, understand the likely energy production and financial implications of a proposed project.

SAM provides comparisons with fossil fuel systems to highlight the benefits and differences of a renewable energy project through various metrics. The model simulates the performance of numerous types of renewable energy projects, which are fine-tuned using geographic-specific weather information. For example, SAM can simulate the expected power that will be generated by a residential PV array in San Diego, California, by incorporating weather conditions in the San Diego area, drawing on a publicly-available database of weather conditions.[17]

SAM also includes optional battery storage, concentrated solar power, solar thermal, wind, geothermal, and biomass power systems. In addition to simulating the performance of a renewable energy project, SAM performs financial analyses based on the results of the simulation to show the implications for projects that either buy and sell electricity at retail rates (residential and commercial), or sell electricity at a price determined in a **power purchase agreement (PPA)**. It calculates the internal rate of return and other key financial metrics, such as the **levelized cost of energy (LCOE)**, electricity cost with and without the renewable energy system, electricity savings, after-tax net present value, and payback period. SAM is able to perform parametric and sensitivity analyses, Monte Carlo simulations, and weather variability studies.[18] Figure 4.7 shows a sample analysis in SAM.

NREL offers yet another useful tool, the **PVWatts® Calculator**, to help estimate the energy production and cost of PV-generated energy throughout the world. It is freely available online at https://pvwatts.nrel.gov/. Owners, consumers, installers, and manufacturers can use this tool to develop performance and cost estimates of potential PV installations. The user inputs the desired location and the DC power rating of the PV array, the array and module types, the retail electricity rate, and tilt and azimuth degrees. The tool generates monthly information about solar radiation, AC energy generated, and calculates the value of that energy. The calculator provides quick projections about potential projects and the user can modify the inputs, such as array and module type, to maximize results. Figure 4.8 shows a sample output of the PVWatts Calculator.

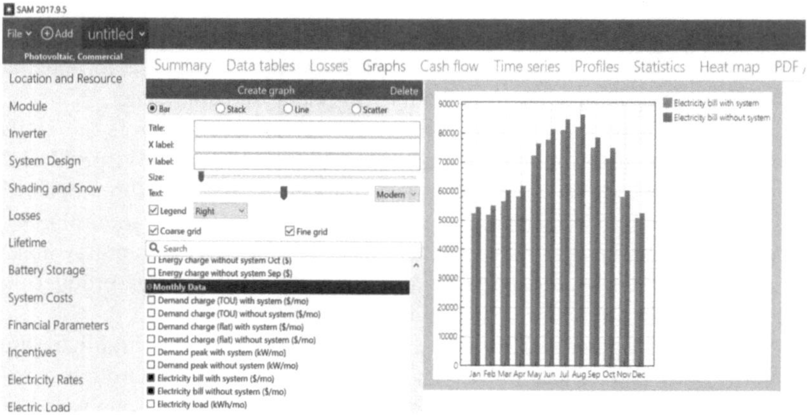

Figure 4.7 Electricity bill with and without PV system for sample data.

Source: SAM Version 2017 9.5.

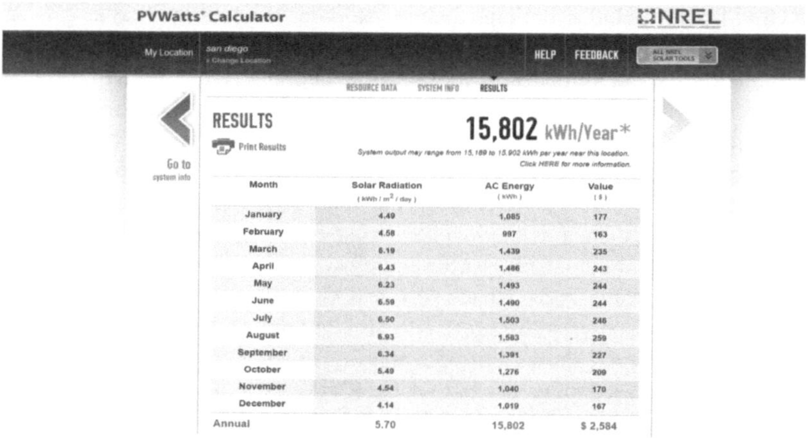

Month	Solar Radiation (kWh / m² / day)	AC Energy (kWh)	Value ($)
January	4.49	1,085	177
February	4.58	997	163
March	5.19	1,439	235
April	6.43	1,486	243
May	6.23	1,493	244
June	6.59	1,490	244
July	6.50	1,503	246
August	5.93	1,583	259
September	6.34	1,391	227
October	5.49	1,276	209
November	4.54	1,040	170
December	4.14	1,019	167
Annual	5.70	15,802	$ 2,584

Figure 4.8 Sample output of PVWatts for a 10 kW PV array in San Diego, California, using standard system settings.

Source: pvwatts.nrel.gov.

Notes

1 Bickel, J.A. "The Basics of Power Monitoring Systems." *Electrical Construction & Maintenance (EC&M) Magazine*. June 05, 2012. Accessed March 24, 2019. www.ecmweb.com/metering-amp-monitoring/basics-power-monitoring-systems.

2 "Tesla App Support." Tesla, Inc. January 26, 2019. Accessed March 24, 2019. www.tesla.com/support/Tesla-app.

3 "How It Works." EnPhase Envoy. Accessed March 24, 2019. https://enphase.com/en-us/homeowners/welcome/how-it-works#envoy.

4 "APsystems Monitoring – APsystems USA | Leading the Industry in Solar Microinverter Technology." APsystems USA | Leading the Industry in Solar Microinverter Technology. Accessed March 24, 2019. https://usa.apsystems.com/products/monitor/.

5 GmbH, Solare Datensysteme. "Log™ Monitoring & Feed-in Management for End-Customers." *Solar*. March 8, 2019. Accessed March 24, 2019. www.solar-log.com/en/end-customers/.

6 "What is Machine Learning? | How It Works, Techniques & Applications." MATLAB & Simulink. Accessed March 24, 2019. www.mathworks.com/discovery/machine-learning.html.

7 Ibid.

8 Perera, K.S., Z. Aung, and W.L. Woon. "Machine Learning Techniques for Supporting Renewable Energy Generation and Integration: A Survey." *Data Analytics for Renewable Energy Integration Lecture Notes in Computer Science* (2014): 81–96. doi: 10.1007/978-3-319-13290-7_7.

9 Risse, M. (2019, March 1). "Using Data Analytics to Improve Operations and Maintenance." *Power Magazine*. Retrieved from www.powermag.com/using-data-analytics-to-improve-operations-and-maintenance/?pagenum=1.

10 Kashyap, S. (2015, June 1). "Predictive Maintenance of Wind and Solar Farms Using Machine Learning Models." *Algo Engines*. Retrieved from http://algoengines.com/2015/06/10/predictive-maintenance-of-wind-and-solar-farms-using-machine-learning-models/.

11 Burger, A. (2017, July 12). "Machine-learning Solar Tracking Technology Nudges PV Field Production Nearer Optimum Levels." *Solar Magazine*. Retrieved from https://solarmagazine.com/machine-learning-solar-tracking-nudges-pv-field-production/.

12 Rudin, C., D. Waltz, R. Anderson, A. Boulanger, A. Salleb, M. Chow, H. Dutta, P. Gross, B. Huang, S. Ierome, F.I. Delfina, A. Kressner, R.J. Passonneau, A. Radeva, and L. Wu. (2011). "Machine Learning for the New York City Power Grid." *IEEE Transactions on Pattern Analysis and Machine Intelligence* 34: 328–45. doi: 10.1109/TPAMI.2011.108. Retrieved from www.researchgate.net/publication/51131062_Machine_Learning_for_the_New_York_City_Power_Grid.

13 "International Performance Measurement and Verification Protocol." Efficiency Valuation Organization. Accessed March 24, 2019. https://evo-world.org/en/library.

14 *M&V Guidelines: Measurement & Verification for Performance Based Contracts, Version 4.0*. Report. US Department of Energy. November 2015. Accessed March 24, 2019. www.energy.gov/sites/prod/files/2016/01/f28/mv_guide_4_0.pdf.

15 "International Performance Measurement and Verification Protocol." Efficiency Valuation Organization. Accessed March 24, 2019. https://evo-world.org/en/library.

16 Ibid.

17 U.S. weather and climactic (historical weather over long time) data can be found online at the National Ocean and Atmospheric Administration's (NOAA) National Weather Service website, www.weather.gov/, and climate data on their Climate Data Online site, www.ncdc.noaa.gov/cdo-web/. For data on the U.K., visit the Met Office at www.metoffice.gov.uk/.

18 U.S. Department of Energy National Renewable Energy Laboratory. (2018, September 10). NREL. Retrieved from System Advisor Model (SAM). https://sam.nrel.gov/.

5 Conclusion

The growing market for PV system inspectors

The long-term trend for U.S. solar jobs looks strong.[1] The U.S. Bureau of Labor Statistics is bullish on jobs in the PV sector, projecting that employment of PV installers will rise 105% from 2016 to 2026, much faster than predicted for the U.S. national average. This expansion will result in excellent job opportunities for qualified individuals, particularly those who have completed PV training courses at a community college or technical school.[2]

With demand for PV inspectors and certifiers closely linked to the growth of the PV sector as a whole, their future looks bright. As discussed in Chapter 1, in 2017 the North American Board of Certified Energy Practitioners (NABCEP) announced that it is now adding the Photovoltaic System Inspector (PVSI) certification to its industry-leading credentials and certifications. In addition, in 2018, NABCEP introduced the PV Commissioning and Maintenance Specialist (PVCMS™) certification.[3]

The rise in demand for PV inspectors/O+M professionals is largely driven by the plethora of state and local building codes. The varying degrees of state and local permitting and inspection requirements play a significant role in PV system uptake.[4] For instance, California, a key market with about 40% of U.S. solar capacity, leads the U.S. with its permitting regime.[5] California has developed a state-wide inspection and certification regime that addresses a variety of concerns, including structural, electrical, and fire prevention requirements.[6]

The new NABCEP credential for PV system inspectors

Recognizing the need for more trained specialists in the operations and maintenance of PV systems, one of NABCEP's newest credentials is the PV System Inspector™ (PVSI). While there are no prerequisites for the exam, comprehensive knowledge of PV systems is assumed. According to the NABCEP Certification Handbook from January 2018,

> Applicants should know how to assess the safety and operation of a system, be able to verify code compliance via interpretation of design plans and building documents, conduct on-site inspections, and report results.[7]

According to the NABCEP Job Task Analysis, a PVSI is "responsible for inspecting residential and commercial photovoltaic systems. They provide inspection services for Authorities Having Jurisdiction (AHJ), utilities, state incentive programs, and financing companies."[8]

A PVSI is knowledgeable in four domains:[9]

I) Inspecting electrical components and systems (44%)

   A. Visual inspection of system labeling
   B. Check components match those on approved plans and meet standards
   C. Inspect conductors and raceways
   D. Inspect conductor terminations
   E. Verify grounding through continuity tests
   F. Inspect point of PV system interconnection with grid
   G. Verify overcurrent protection devices

II) Inspecting energy storage components and systems (21%)

   A. Verify PV array and design control
   B. Verify energy storage system design and installation
   C. Check multi-mode inverter
   D. Check AC connections

III) Inspecting mechanical/structural components and systems (21%)

   A. Inspect roof-mounted components and systems
   B. Inspect ground-mounted components and systems

IV) Documentation for the system inspection (14%)

   A. Review permit package for completeness, accuracy and compliance
   B. Quantify and report deviations from design documentation

Knowledge of the following areas is required:

- PV system components
- AC power distribution system
- Electrical safety
- Grounding and bonding
- Ground fault hazards
- Circuit current calculations
- Conductor ampacity calculations
- Conductor coloring
- Conduit fill calculations

- Expansion joint calculations
- Raceway support code compliance
- NEMA ratings
- Working clearances
- AHJ and utility labeling requirements
- National Electrical Code, in particular:
    - Article 690 photovoltaic systems
    - Article 691 large scale PV electric power production facility
    - Article 705 interconnected electric power production sources
    - Article 706 energy storage systems
- Safety practices
- Corrosive properties of metals
- Reading blueprints
- Electrical schematics and symbols
- String sizing calculations
- Safety practices
- Series and parallel DC circuits
- Energy storage system inverter power flow
- Battery technologies and storage types
- Overcurrent protection device sizing calculations

The Job Task Analysis for the PV Commissioning and Maintenance Specialist (PVCMS) includes the information shown in Table 5.1.

Table 5.1 Job Task Analysis for the PV Commissioning and Maintenance Specialist (PVCMS)[10]

Description	*Percent of test*
Review or develop commissioning protocol	10%
Complete visual and mechanical inspection	12%
Conduct mechanical tests	8%
Conduct electrical tests	15%
Verify system operation	10%
Confirm project completion	8%
Orient end user to system	8%
Verify system operation and performance	8%
Perform preventive maintenance	10%
Perform corrective maintenance	10%

For both tests, knowledge of the following areas is required:

- Standards
  - IEC 62446–1 photovoltaic (PV) systems – Requirements for testing, documentation and maintenance – Part 1: grid connected systems – documentation, commissioning tests and inspection[11]
  - NFPA 70E (National Fire Protection Association – Standard for Electrical Safety in the Workplace)
    - Lock Out Tag Out
- Electric service provider requirements
  - Disconnecting means, interconnection method, equipment location, labeling requirements
- Use of diagnostic tools
  - Multimeter, insulation resistance tester
- Tests
  - Field testing
  - Witness testing
  - Polarity testing
  - DC string open circuit voltage (Voc) testing
  - DC string maximum power current (lmp) testing
  - DC string short circuit (Isc) testing
  - Continuity testing
  - Ground resistance testing
  - AC voltage testing
  - Thermal imaging
  - Wire termination torque verification
  - Torque
- Failure modes for:
  - Modules
  - Mounting/racking
  - Wiring
  - Batteries
- System cleaning
- System monitoring, controls, communications platforms, and performance analytics
- Turnover documentation

The following questions represent the range of material on the test. However, previous experience and knowledge in the NEC is required for the exam. See the Further Resources section at the end of the chapter for additional study materials. In particular, *PV and the NEC* by Bill Brooks and Sean White provides an excellent overview of PV-related content in the NEC.[12]

Sample exam questions

1. When resting a ladder on the side of a building, what is the optimal angle from the ground to the ladder?
2. Given the following information, determine the maximum number of modules that can be placed in a string.

   Coldest expected temperature = −15° C
   Temperature coefficient Voc (open circuit voltage) = −0.33%/C
   Voc of module = 39V
   Max input voltage of inverter = 600V

3. An array has 3 strings of 7 modules. The 3 strings are wired as parallel source circuits. The maximum powerpoint current of one module is 4.9 amps and the short circuit amperage of one module is 5.2 amps. Calculate the maximum current for wire sizing, which should be labeled on a DC disconnect as Imax (maximum current).
4. Which one of the following single conductor cable types can be installed in exposed locations in PV source circuits?

   a THHN
   b THHWN
   c USE-2
   d XHHW

5. How should rust be prevented on battery terminals?

   a Place batteries in ventilated locations
   b Use antioxidant material
   c Keep batteries guarded
   d Treat with acidic solution

Sample question answers

1 The optimal angle from the ground is 75 degrees. For every 4 feet of height, set the ladder out 1 foot. The resulting angle will be about 75 degrees.

2 Coldest expected temperature = −15° C

Difference between coldest temperature and 25° C standard test conditions (STC) = (−15 − 25) = −40

Temperature coefficient Voc (open circuit voltage) =

– 40° C × −0.33%/C (from module specifications) = 13.2% increase Voc

Voc of module = 39V

Increase in Voc cold = 13.2/100 + 1 = 1.132 × 39V = 44.15 Voc

Max input voltage of inverter = 600V

Max number of modules in series = 600V/44.15 Voc cold = 13.59 modules

ROUND DOWN = 13 modules

3 Take the short circuit current and multiply it by the number of parallel connections. Then, multiply it by 1.25 (125%). The National Electrical Code (NEC) 690.8(A)(1)(1) requires that 125% of short circuit current be used because PV current is wild as it is influenced by sunlight. An additional 125% is used for continuous current in sizing overcurrent protection devices. This is 1.56 × the parallel module's short circuit current.

5.2 amps per string × 3 strings × 1.25 = 19.5 amps

Note: Recognizing that module level power electronics can control current, in the 2017 NEC, the "engineering supervision method" was introduced as an alternative approach under 690.8(A)(1)(2). This method requires a licensed Professional Engineer (PE) to use an industry standard method.

4 c – USE-2
USE-2 or PV wire must be used because it is sunlight resistant, rated for wet conditions, and can handle temperatures up to 90° C.

5 b – use antioxidant material
Antioxidants are chemicals that form a passivation layer around metals that inhibit oxidation on certain metals. Oxidation occurs when oxygen in the air reacts with a metal, such as steel, to form iron oxide, or rust.

The future of O+M and M+V in the solar PV industry

The move toward reliability-centered maintenance represents a more nuanced understanding of our equipment. Also, unlike any time before, we now have tremendous amounts of data from our systems. Indeed, we live in an era of data *overload*. The next revolution in maintenance will be the ability to effectively use the data to improve and prolong our systems.

Machine learning will allow us to accurately predict array performance anywhere in the world, and to solve difficult-to-diagnose issues, such as ground faults and inverter gating failure. As the cost of robotics declines, array washing with robots and self-cleaning modules will rise. For instance, in a DOE-funded project, Boston University developed a waterless cleaning solution using a transparent electrodynamic screen that lifts dust particles from a concentrated solar collector through traveling-wave electric fields. In lab tests, the screen removed over 90% of dust in minutes using a small amount of the collector's own energy.[13]

High speed non-contact scanners are facilitating the detection of hot spots in PV modules, and aerial drones are aiding in visual and thermographic inspections. Breakthroughs in inverter life extension are on the horizon. Improvements include new circuit designs that improve packaging, fewer parts, and wide-bandgap semiconductor switches that can operate at high temperatures. Solar LCOE would decline dramatically if inverters did not need to be replaced during the system's lifetime.[14]

It is an exciting time to be in solar energy. Diving into solar provides you with the opportunity to be part of a rapidly-growing industry, and part of the solution to global environmental crises. Whatever your role, from an investor to a facility manager to a technician responsible for installation and maintenance of systems, you will play a vital role in providing clean, reliable, and low-cost energy to millions.

I hope that the knowledge you have gained from this book has shown you new pathways to success and insight into how solar PV works. So what are you waiting for? Time to join the solar revolution!

Further resources

North American Board of Certified Energy Practitioners (NABCEP) test preparation

White, Sean. *Solar Photovoltaic Basics: A Study Guide for the NABCEP Entry Level Exam 2nd Edition*. Earthscan Routledge. August 29, 2018.

White, Sean. *Solar PV Engineering and Installation: Preparation for the NABCEP PV Installation Professional Certification*. 1st Edition, Earthscan Routledge. May 15, 2018.

National Electrical Code

Brooks, Bill, and White, Sean. *Photovoltaic Systems and the National Electric Code*. 1st Edition. Routledge. March 24, 2018.

Holt, Mike. *Mike Holt's Illustrated Guide to Understanding NEC Requirements for Solar Photovoltaic Systems*. Based on the 2017 NEC. Mike Holt Entreprises. www.mikeholt.com/productitem.php?id=1476.

Solar photovoltaic science and systems integration

Ginsberg, Michael. *Harness It: Renewable Energy Technologies and Project Development Models Transforming the Grid*. Business Expert Press. May 2019. www.businessexpertpress.com/books/harness-it-renewable-energy-technologies-and-project-development-models-transforming-the-grid/.

Honsberg, Christiana, and Bowden, Stuart. PVEducation (online textbook). www.pveducation.org/.

Fthenakis, Vasilis, and Lynn, Paul A. *Electricity from Sunlight: Photovoltaic-Systems Integration and Sustainability*. 2nd Edition. Wiley. March 2018.

Parrish, Peter. *Photovoltaic Laboratory: Safety, Code-Compliance, and Commercial Off-the-Shelf Equipment 1st Edition*. CRC Press. April 7, 2016.

Notes

1 *2017 National Solar Jobs Census*. The Solar Foundation. November 2017.

2 *Occupational Outlook Handbook: Solar Photovoltaic Installers*. Publication. Bureau of Labor Statistics. Bureau of Labor Statistics, 2018. Accessed March 25, 2019. Retrieved from www.bls.gov/ooh/construction-and-extraction/solar-photovoltaic-installers.htm#tab-6. July 25, 2018.

3 North American Board of Certified Energy Practitioners. “Press Release: PV System Inspector and Solar Heating System Inspector Certifications.” News Release, North American Board of Certified Energy Practitioners. Accessed March 25, 2019. www.nabcep.org/news/pv-system-inspector-and-solar-heating-system-inspector-certifications. May 30, 2017.

4 *2017 National Solar Jobs Census*. The Solar Foundation. November 2017. p. 49.

5 Ibid., p. 4.

6 California Governor's Office. “California Solar Permitting Guidebook.” Accessed March 25, 2019. http://opr.ca.gov/docs/California_Solar_Permitting_Guidebook.pdf. 2012.

7 Handbook. North American Board of Certified Energy Practitioners, Inc. (NABCEP®). Accessed March 25, 2019. www.nabcep.org/wp-content/uploads/2018/02/NABCEP-Certification-Handbook-V2018.compressed.pdf. January 2018.

8 Photovoltaic System Inspector Job Task Analysis. www.nabcep.org/wp-content/uploads/2017/05/NABCEP-PVSI-JTA_17.pdf. April 2017.

9 Note the sub-tasks are paraphrased from the Job Task Analysis.

10 *Photovoltaic Specialists: Job Task Analysis*. Report. North American Board of Certified Energy Practitioners, Inc. (NABCEP®). Accessed March 25, 2019. www.nabcep.org/wp-content/uploads/2018/07/NABCEP-PV-Specialist-JTA-Rev_7-27-18.pdf. September 2017.

11 "IEC 62446-1: Photovoltaic (PV) Systems – Requirements for Testing, Documentation and Maintenance – Part 1: Grid Connected Systems – Documentation, Commissioning Tests and Inspection." IHS Market Standards Store. Accessed March 25, 2019. https://global.ihs.com/doc_detail.cfm?document_name=IEC 62446-1&item_s_key=00668699. August 2018.
12 Brooks, B., and S. White. *Photovoltaic Systems and the National Electric Code*. Abingdon, Oxon: Routledge, 2018.
13 "Self-Cleaning CSP Collectors." *Energy.gov*. Accessed March 25, 2019. www.energy.gov/eere/solar/downloads/self-cleaning-csp-collectors.
14 "Extending Solar Energy System Lifetime with Power Electronics." *Energy.gov*. Accessed March 25, 2019. www.energy.gov/eere/solar/articles/extending-solar-energy-system-lifetime-power-electronics.

Glossary

Preface and Chapter 1

Authorities Having Jurisdiction (AHJs) The organization within local government responsible for granting permits and inspecting PV systems and other modifications to a building.

measurement and verification (M+V) Planning, measuring, collecting, and analyzing data to ensure savings are achieved as a result of an investment.

net present value (NPV) The value in today's dollars of all future savings from an investment minus the present cost of the investment.

North American Board of Certified Energy Practitioners (NABCEP) Professional certification organization that offers individual and company accreditation for PV system installers and other types of renewable energy professionals.

operations and maintenance (O+M) A program for the regular upkeep of facilities.

photovoltaic (PV) The production of electricity from sunlight.

PV System Inspector (PVSI) Credential through the North American Board of Certified Energy Practitioners to be recognized as an expert in the inspection of PV systems.

Chapter 2

alternating current (AC) Electrons "jiggle" back and forth in a sinusoidal (sine) wave, and the electrons at the source are not the ones used by the load.

Balance of System (BoS) Network of components required to condition and transmit the electricity from PV cells. This includes inverters, wires, circuit breakers, monitoring equipment and software, and sometimes charge controllers, batteries, and transformers.

bypass diode A device that allows current to flow in only one direction. Bypass diodes are connected in parallel either with an individual solar cell, ideally, or series of cells, to be economical. Under normal operation, the bypass diodes have no effect on the circuit because they have opposite polarity, or are reverse biased.

capacitor Device that stores energy and aids in the switching of a transistor by ensuring the appropriate amount of energy is output to the transistor. Also used in electrical grids to increase the voltage during peak demand.

cell (or module) mismatch Differences in cell or module characteristics (current, voltage) that lead to reduced output when connected in series. The lowest current determines the current of the string.

charge controller Regulates the current (amperage) from the PV system to the battery to ensure the battery does not become over-charged or over-drained.

conduction band The band where electrons jump to from the valence shell. When in the conduction band, the electrons have enough energy to move freely in the material.

conductor Material that allows the flow of electricity. There is no energy bandgap for electrons to move from the valence shell to the conduction band, and thus no electric field can be established. Common conductors include copper and aluminum.

covalent bond Sharing of electron pairs between atoms.

current The rate of flow of electric charge.

direct current (DC) Electrons travel directly from the voltage source to the load. The same electrons generated from the source are used by the load. Note that batteries and solar cells produce DC.

electricity Using moving electrons to power loads.

energy Work being done (Joule, J). The force of lifting an apple (one Newton) one meter is one Joule. It is power times time.

energy bandgap The energy required for valence shell electrons to move into the conduction band. Conductors' valence and conduction bands overlap, so there is no energy bandgap. Semiconductors have a small energy bandgap, while insulators have a large bandgap that is insurmountable. The energy bandgap determines the absorption properties of the semiconductor. A material absorbs only those photons whose energies equal or exceed the bandgap. Material with a high energy bandgap absorbs short wavelength photons, while material with a low energy bandgap absorbs long and short wavelength photons, but energy in excess of the bandgap (short wavelength) is wasted as thermal energy.

hot spot Part of a module where overheating is occurring due to a shaded or short-circuited cell. Reduces module output and leads to degradation.

inverter Device that converts DC to AC and conditions electricity to match the requirements of the load or the electrical grid along the parameters of voltage, phase, frequency, and waveform. In addition, inverters track and capture electricity from the solar array at its maximum voltage and current, called the maximum powerpoint (MPP).

lithium ion (Li+) battery Stores electrical energy in the form of chemical potential energy, and discharges to electrical circuit. Lithium ions are transferred between electrodes. Compared to other batteries, they are light and have high energy density and a high depth of discharge. For this reason they are optimal for PV systems.

load A component in an electrical circuit that consumes power. Utilities refer to their customers as loads.

maximum powerpoint tracking (MPPT) How inverters optimize the output of an array. In MPPT inverters, power is obtained from the array at the point at which the greatest power is generated. The greatest power is obtained where the product of voltage and current is highest since power equals voltage × current.

Ohm's Law A way to determine the magnitudes of different components of electricity. For direct current, voltage equals current times resistance.

Photoelectron An electron that absorbs the energy of a photon and is dislodged from the atom

photoelectric effect The production of electricity through sunlight. Photons energize electrons in the valence shell of semiconductors into the conduction band. These electrons are then used in a circuit.

photon A wave made from a stream of quantum particles, or packets of energy from the sun.

power The rate at which work is done (watt, W). Power equals voltage times current. One watt equals one Joule divided by one second.

PV cell A semiconductor that converts photons into direct current electricity.

resistance The degree to which material restricts electrical flow (Ohm, Ω). Resistance equals voltage divided by current.

semiconductor Material that conducts electricity but offers opposition to current flow. The energy bandgap is relatively small in size (around 1 eV) so that some electrons can be excited from the valence shell to the conduction band. Unlike conductors, semiconductors of different chemical compositions can be placed together to create the electric field necessary to produce PV power. Common semiconductors include silicon, carbon, and germanium.

shunt Alternate path for current flow with lower resistance than the desired path. In solar cells, shunts occur at the cell edges and are due to impurities near the edges that provide a short circuit path around the p-n junction, leading to decreased performance.

standard test conditions (STC) Conditions at which PV modules are tested and upon which ratings are based. Modules are tested at a cell temperature of 25° C, an irradiance of 1,000 watts per square meter, and an air mass of 1.5.

transistor Used in inverters and many electronic devices as a switch and signal amplifier.

transformer Electrical device that transforms power through induction only (except in the case of an autotransformer) to increase or decrease voltage, and proportionally decrease or increase current. A step-up transformer increases voltage, while a step-down transformer decreases voltage.

transformerless inverter Inverter without a transformer. Uses a computerized multistep process instead of a transformer to invert from DC to AC. This reduces the weight and increases the efficiency of the inverter, although it has been slow to gain acceptance in the U.S. due to the lack of electrical isolation between the DC and AC sides of the circuit, which were traditionally isolated by transformers.

valence shell The outer shell of an atom. When electrons in the valence shell gain the energy from a photon or any outside force, such as a charge, they will break away from the atom and become free electrons.

voltage Electromotive force or potential (Volt, V). Voltage equals current times resistance.

voltage source The source of power in a circuit (i.e. a battery or generator).

voltaic Production of electricity.

Chapter 3

arc flash The light and heat produced due to medium to high voltage electricity traveling through the air between conductors or from a conductor to ground. Specialized PPE is needed for electricians working within the arc flash boundary on energized systems of such voltages. An arc flash can occur at any voltage above 120 volts, although are more common with medium voltages of 600 volts and above.

DC ground fault Unintended connection between a current-carrying conductor and an equipment grounding conductor or any metallic component that is grounded. This is a potentially dangerous condition that could result in equipment damage and fire.

failure mode and effects analysis (FMEA) A procedure for identifying and prioritizing all possible failures in a product, system, or design. A risk priority number is assessed to each failure mode based on the severity, probability, and detectability of the potential failure.

Lock Out Tag Out (LOTO) One of the most important work practices for technicians, LOTO is a way for workers to protect themselves and others from potentially hazardous energy and communicate with others the work that is being performed. As the name implies, the worker locks out a system and then tags out that system. A lock ensures that whatever is being worked on is off (de-energized – i.e. a circuit breaker is open). A tag is used to communicate with technicians and others who is performing the work and what work is being performed.

mean time between failures The predicted number of hours between one failure and the next.

overcurrent protection devices Tools that protect equipment and people from current that is greater than the rated current of the equipment or conductors as a result of a short circuit, ground fault, or overload. Types include fuses, circuit breakers, and ground fault circuit interrupters.

personal protective equipment (PPE) Protective clothing that serves as a last protection against workplace hazards. Includes helmets, glasses/goggles, boots, gloves, and ear protection.

preventive maintenance (PM) Regularly-scheduled repairs that have been shown to reduce equipment failure and unplanned emergency repair time that would otherwise be incurred by relying on reactive maintenance (waiting until something breaks to fix it). Another benefit of PM is that it reduces planned repair hours.

procedures/work practices Following a risk assessment, these are the control measures followed to minimize risks to technicians and other individuals.

reactive maintenance Waiting until something breaks to fix it.

reliability-centered maintenance (RCM) Equipment-specific O+M plan that takes into account equipment criticality levels, the impact of a failure, and potential failure modes.

time current curve A graph on a circuit breaker that shows how quickly it will trip. Time to trip decreases with time logarithmically.

Chapter 4

derivative gain When coupled with proportional and integral gain, this algorithm in a programmable logic controller accounts for the rate of change of an error, and a setpoint is achieved with less overshoot and oscillations.

energy data system (EDS) Collection of data (inputs) from sensors into a computer.

energy information system (EIS) Collection and display of data on a terminal from sensors so a facility manager can analyze data and decisions made that impact a facility's operations and financials.

energy management system (EMS) In addition to the collection and display of data, an energy management system allows facility managers and technicians to automatically and manually control their equipment through programmable logic controllers.

graphical user interface (GUI) or **human machine interface (HMI)** Terminal (screen) for a person to access and interact with data.

integral gain When coupled with proportional gain, this algorithm on the programmable logic controller increases the movement towards the setpoint by accounting for the duration of past errors.

International Performance Measurement and Verification Protocol (IPMVP) Protocol for measurement and verification to ensure the accurate assessment of performance and financial metrics for energy efficiency and renewable energy projects. Run by the Efficiency Valuation Organization (EVO). https://evo-world.org/en/products-services-mainmenu-en/protocols/ipmvp.

levelized cost of energy (LCOE) The cost of a power system when accounting for lifetime costs in addition to upfront cost (in cents per kWh). Used to accurately compare costs across different power sources.

machine learning (ML) An algorithm that allows a computer to learn from data without explicit instructions.

microgrid Interconnected loads and distributed energy resources that can operate both connected to, and isolated from, the central grid – for instance, when PV systems are connected with batteries and other energy sources, such as wind turbines, combined cycle power plants, and diesel generators.

power purchase agreement (PPA) A contract between a power generator and a buyer to purchase electricity at an agreed-upon price in cents per kWh over the lifetime of the power plant. The buyer is typically a utility that operates the transmission and distribution lines for a region.

programmable logic controller (PLC) A computer that controls and coordinates the operation of equipment. The PLC is at the top of a hierarchy of system and terminal unit controllers, coordinating their operations.

proportional gain An algorithm in a PLC that corrects errors proportional to the size of the difference between the setpoint and the current value (or control point) multiplied by an adjustable constant. The greater the constant, the quicker the response, but this leads to overshoot. The overshoot is corrected by integral and derivative gain.

proportional, integral, derivative gain (PID) controller A controller that uses an algorithm for optimizing the performance of equipment, such as motors, to achieve and stay within desired settings, such as for temperature.

PVWatts® Calculator A calculator developed by the National Renewable Energy Laboratory (NREL) that estimates energy production and the cost of PV generated energy throughout the world. It is freely available online at https://pvwatts.nrel.gov/.

supervised learning A subset of machine learning that uses classification and regression to develop predictive models based on training data.

System Advisor Model (SAM) A software developed by the National Renewable Energy Laboratory (NREL) that calculates the performance and financial metrics of renewable energy projects. It is freely available online at https://sam.nrel.gov/.

training data Known information upon which future predictions are made in supervised learning.

unsupervised learning A subset of machine learning in which the algorithm has no training data, and thus no prior known information. It identifies patterns within the data. Cluster analysis, in which data is grouped based on similarities, is one of the most common techniques and is often used for image analysis and pattern recognition.

Index

Note: Page numbers in **bold** indicate a table and page numbers in *italics* indicate a figure on the corresponding page.

Adams, William Grylls 1
alternating current (AC) 26–27
arc flash personal protective equipment 73, *74*, *75*
array washing 87, 88–89
azimuth 29–30

Balance of System 39–41, **42**
battery 50–53, *51*, **52**
Becquerel, Alexandre Edmond 1
Berman, Elliot 2
bypass diodes 43

California 131
camera, infrared 83, *84*
capacitor 45
capacitor failure **91**, *93*, **95**
capacity, installed **5**
cell mismatch 43
Chapin, Daryl 1
charge controller 50, **99**, **107**
chemistry, solar cell 36–38, *37*
circuit *21*; analysis 22–23; parallel 24–25, *25*; series 23, *23*
circuit breaker 63, *64*, **102**, 105, *105*, **108**
cluster analysis 125
cold temperature sizing 30–32, **32**
conductor 20
current 22; alternating 26–27; direct 26–27

Day, Richard Evans 1
DC ground fault 49, **90**, **94**
degradation rates 83
dehydration **57**
depth of discharge (DoD) 51
derivative gain 123
diodes, bypass 43
direct current (DC) 26–27

Edison, Thomas 19, 26
EDS *see* energy data systems (EDS)
EGC *see* equipment grounding conductor (EGC)
Einstein, Albert 1
electricity: basics 19–27; components 19–20, *20*, *21*
electrocution 55, **56**, 58, 59–65
employment *6*, 6–7
energy, power *vs.* 21–22
energy data systems (EDS) 119
energy density 52
Energy Monitoring and Analysis (EMA) (software) 121
EnPhase Envoy (software) 121
equipment grounding conductor (EGC) 63
eye protection 71, *71*

face shield 73
failure mode and effects analysis (FMEA) 76–79, *77–78*, **80**
falls 55, **57**, 68, 71
"fatal four" 55
flow battery **52**
FMEA *see* failure mode and effects analysis (FMEA)

footwear 71, *72*
forward bias 43
Fuller, Calvin S. 1
fuse 63, *64*

gating fault *93*, **96**
GEC *see* grounding electrode conductor (GEC)
GFCIs *see* ground fault circuit interrupters (GFCIs)
GFDIs see ground fault detection interrupters (GFDIs)
graphical user interface (GUI) 119, 120
grid parity *3*
ground fault circuit interrupters (GFCIs) 63, *64*
ground fault detection interrupters (GFDIs) **90**, **102**
grounding electrode conductor (GEC) 62–63, *63*
guard rails 66, *67*
GUI *see* graphical user interface (GUI)

heat stroke **57**
high shunt resistance 44–45
HMI *see* human machine interface (HMI)
hot spot 43
human machine interface (HMI) 118, 120
hypothermia **57**

IGBT *see* insulted gate bipolar transistor (IGBT)
infrared camera 83, *84*
installed capacity **5**
insulated gate bipolar transistor (IGBT) 48
integral gain 123
International Performance Measurement and Verification Protocol (IPMVP) 126
inverters 45–49
IPMVP *see* International Performance Measurement and Verification Protocol (IPMVP)

jobs, solar *6*, 6–7

ladder safety 68, *68*, *69*
latitude 29–30
latitude tilt 29
LCOE *see* levelized cost of energy (LCOE)
lead acid battery **52**
legend: likelihood of detection *78*; probability 77; severity 77
levelized cost of energy (LCOE) 127, 137
lithium ion battery **52**
lithium plating **97**, **99**, *100*
Lock Out Tag Out (LOTO) 59, *60*, 61
low circuit series resistance 45

machine learning (ML) 124–125
maintenance 76–116
manufacturing *2*, 2–6
maximum powerpoint (MPP) **42**, 48, *48*, 49, 50
measurement 125–128
metal oxide semiconductor field effect transistor (MOSFET) 47–48
meters 109–110, *109*, *110*, *111*, *112*
Millikan, Robert 1
ML *see* machine learning (ML)
MOSFET *see* metal oxide semiconductor field effect transistor (MOSFET)

North American Board of Certified Energy Practitioners (NABCEP) 17

OCPDs *see* overcurrent protection devices (OCPDs)
Ohm's Law 22, *23*
OSHA 69, 71
overcurrent protection devices (OCPDs) 62

Pearson, Gerald 1
personal fall arrest system (PFAS) 71–72, *73*
personal protective equipment (PPE) 55, 69–73
PFAS *see* personal fall arrest system (PFAS)
photoelectric effect 33
photoelectron 33
photon 33
physics, of solar cell 33–36
PID controller 123, *124*
Planck, Max 35

PLC controller 122
PM *see* preventive maintenance (PM)
power, energy *vs.* 21–22
power purchase agreement (PPA) 127
Power Triangle 24, *25*
PPA *see* power purchase agreement (PPA)
PPE *see* personal protective equipment (PPE)
predictive maintenance *76*
preventive maintenance (PM) 76, *76*
proactive maintenance *76*
proportional gain 122
PV Commissioning and Maintenance Specialist (PVCMS) **133**, 133–135
PV System Inspector™ (PVSI) 131–133
PVWatts 127
pyranometer 27

racking **102**, **107**
Railway Method 25–26
rapid shutdown 65
RCM *see* reliability-centered maintenance (RCM)
reactive maintenance 76, *76*
reliability-centered maintenance (RCM) 76, *76*
resistance 20; high shunt 44–45; low circuit series 45
reverse bias 43
risk assessment 55–58, *55*, **56–57**, *58*
roundtrip efficiency 52

safety barriers 65–66
safety glasses 71, *71*
safety nets 67, *67*
safety watch 59
SAM *see* System Advisor Model (SAM)
sample exam questions 135
semiconductors 33
shunts 44
silicon *37*
solar array components 43–53
solar azimuth angle 30
solar cell: chemistry 36–38, *37*; physics 33–36
solar energy basics 27–41
solar irradiation, global map *28*
Solar-Log (software) 121
solar module 39–41, *40*
solar photovoltaic industry, history of 1–6
string inverters, cold temperature sizing for 30–32, **32**
supervised learning 125
System Advisor Model (SAM) 127
system size 7–17

templates, FMEA and maintenance schedule **113–116**
Tesla, Nicola 19, 26
thermal runaway *93*, **97**, **99**, 100
thermography 83, *84*, **85**
thyristor 48
time current curve (TCC) *105*, 105–106
tools **103**
tracker **103**
training data 125
transformer **104**, **108**
transformerless inverters 49
transistor 45–48, *46*, *47*

unit conversion 25–26
unsupervised learning 125

valence 33
verification 125–128
Volta, Alessandro 20
voltage 22
voltage source 20

washing array 87, 88–89
wiring **95**, **107**